SOLAR RADIATION

Practical Modeling for
Renewable Energy Applications

ENERGY AND THE ENVIRONMENT

SERIES EDITOR
Abbas Ghassemi
New Mexico State University

PUBLISHED TITLES

Solar Radiation: Practical Modeling for Renewable Energy Applications
Daryl R. Myers

Solar and Infrared Radiation Measurements
Frank Vignola, Joseph Michalsky, and Thomas Stoffel

Forest-Based Biomass Energy: Concepts and Applications
Frank Spellman

Introduction to Renewable Energy
Vaughn Nelson

Geothermal Energy: Renewable Energy and the Environment
William E. Glassley

Solar Energy: Renewable Energy and the Environment
Robert Foster, Majid Ghassemi, Alma Cota
Jeanette Moore, and Vaughn Nelson

Wind Energy: Renewable Energy and the Environment
Vaughn Nelson

SOLAR RADIATION
Practical Modeling for
Renewable Energy Applications

Daryl R. Myers

CRC Press
Taylor & Francis Group
Boca Raton London New York

CRC Press is an imprint of the
Taylor & Francis Group, an **informa** business

CRC Press
Taylor & Francis Group
6000 Broken Sound Parkway NW, Suite 300
Boca Raton, FL 33487-2742

First issued in paperback 2017

© 2013 by Taylor & Francis Group, LLC
CRC Press is an imprint of Taylor & Francis Group, an Informa business

No claim to original U.S. Government works

Version Date: 20121218

ISBN 13: 978-1-138-07554-2 (pbk)
ISBN 13: 978-1-4665-0294-9 (hbk)

Library of Congress Cataloging-in-Publication Data

Myers, Daryl.
    Solar radiation : practical modeling for renewable energy applications / Daryl Ronald Myers.
       pages cm. -- (Energy and the environment)
    Includes bibliographical references and index.
    ISBN 978-1-4665-0294-9
    1. Solar radiation--Measurement. 2. Solar radiation--Mathematical models. 3. Solar energy--Estimates. I. Title.

QC912.M49 2013
551.5'27101151--dc23   2012037358

Visit the Taylor & Francis Web site at
http://www.taylorandfrancis.com

and the CRC Press Web site at
http://www.crcpress.com

This book is dedicated to the next generation of solar radiation resource and renewable energy scientists and engineers who will undoubtedly improve greatly on the methods and results presented here.

Contents

Series Editor

Dr. Abbas Ghassemi is the director at the Institute for Energy and Environment (IE&E) and professor of chemical engineering at New Mexico State University. He oversees the operations of WERC: A Consortium for Environmental Education and Technology Development, the Southwest Technology Development Institute (SWTDI), and the Carlsbad Environmental Monitoring and Research Center (CERMC). As the director of IE&E, Dr. Ghassemi is also the chief operating officer for programs worth over $10 million annually in education and research and outreach in energy resources, including renewable energy, water quality and quantity, and environmental issues. He is responsible for administrative duties, operation, budget, planning, and personnel supervision for the program. Dr. Ghassemi has authored and edited several textbooks and has many publications and papers in the areas of energy, water, waste management, process control, thermodynamics, transport phenomena, education management, and innovative teaching methods. His research areas of interest include risk-based decision making, renewable energy and water, energy efficiency and pollution prevention, multiphase flow, and process control. Dr. Ghassemi holds a BS, an MS, and a PhD in chemical engineering with minors in mathematics and experimental statistics from the University of Oklahoma and New Mexico State University, respectively, and serves on a number of public and private boards, editorial boards, and peer review panels.

.

Series Editor's Preface

By 2050, the demand for energy could double, or even triple, as the global population rises and developing countries expand their economies. All life on earth depends on energy and the cycling of carbon. Energy is essential for economic and social development but also poses an environmental challenge. We must explore all aspects of energy production and consumption, including energy efficiency, clean energy, global carbon cycle, carbon sources and sinks, and biomass as well as their relationship to climate and natural resource issues. Knowledge of energy has allowed humans to flourish in numbers unimaginable to our ancestors. The world's dependence on fossil fuels began approximately 200 years ago. Are we running out of oil? No, but we are certainly running out of the affordable oil that has powered the world economy since the 1950s. We know how to recover fossil fuels and harvest their energy for operating power plants, planes, trains, and automobiles, which results in modifying the carbon cycle and additional greenhouse gas emissions. This has created the debate on availability of fossil energy resources, peak oil era and timing for the anticipated end of the fossil fuel era, and price and environmental impact versus various renewable resources and use, carbon footprint, emission, and control, including cap and trade and emergence of "green power."

Our current consumption has largely relied on oil for mobile applications and coal, natural gas, and nuclear or water power for stationary applications. To address the energy issues in a comprehensive manner, it is vital to consider the complexity of energy. Any energy resource—including oil, coal, wind, biomass, and so on—is an element of a complex supply chain and must be considered in the entirety as a system from production through consumption. All of the elements of the system are interrelated and interdependent. Oil, for example, requires consideration for interlinking of all of the elements, including exploration, drilling, production, water, transportation, refining, refinery products and by-products, waste, environmental impact, distribution, consumption/application, and finally emissions. Inefficiencies in any part of the system will have an impact on the overall system, and disruption in any one of these elements would cause major interruption in consumption. As we have experienced in the past, interrupted exploration will result in disruption in production, restricted refining and distribution, and consumption shortages; therefore, any proposed energy solution requires careful evaluation and, as such, may be one of the key barriers to implement the proposed use of hydrogen as a mobile fuel.

Even though an admirable level of effort has gone into improving the efficiency of fuel sources for delivery of energy, we are faced with severe challenges on many fronts. This includes population growth, emerging economies, new and expanded usage, and limited natural resources. All energy solutions include some level of risk, including technology snafus, changes in market demand, economic drivers, and others. This is particularly true when proposing energy solutions involving implementation of untested alternative energy technologies.

There are concerns that emissions from fossil fuels will lead to climate change with possible disastrous consequences. Over the past five decades, the world's collective greenhouse gas emissions have increased significantly even as efficiency has increased, resulting in extending energy benefits to more of the population. Many propose that we improve the efficiency of energy use and conserve resources to lessen greenhouse gas emissions and avoid a climate catastrophe. Using fossil fuels more efficiently has not reduced overall greenhouse gas emissions due to various reasons, and it is unlikely that such initiatives will have a perceptible effect on atmospheric greenhouse gas content. While there is a debatable correlation between energy use and greenhouse gas emissions, there are effective means to produce energy, even from fossil fuels, while controlling emissions.

There are also emerging technologies and engineered alternatives that will actually manage the makeup of the atmosphere but will require significant understanding and careful use of energy. We need to step back and reconsider our role and knowledge of energy use. The traditional approach of micromanagement of greenhouse gas emissions is not feasible or functional over a long period of time. More assertive methods to influence the carbon cycle are needed and will be emerging in the coming years. Modifications to the cycle mean we must look at all options in managing atmospheric greenhouse gases, including various ways to produce, consume, and deal with energy. We need to be willing to face reality and search in earnest for alternative energy solutions. There appears to be technologies that could assist; however, they may not all be viable. The proposed solutions must not be in terms of a "quick approach" but a more comprehensive, long-term (10, 25, and 50 plus years) approach that is science based and utilizes aggressive research and development. The proposed solutions must be capable of being retrofitted into our existing energy chain. In the meantime, we must continually seek to increase the efficiency of converting energy into heat and power.

One of the best ways to define sustainable development is through long-term, affordable availability of resources, including energy. There are many potential constraints to sustainable development. Foremost is the competition for water use in energy production, manufacturing, farming, and others versus a shortage of freshwater for consumption and development. Sustainable development is also dependent on the earth's limited amount of soil, and in the not too distant future, we will have to restore and build soil as a part of sustainable development. Hence, possible solutions must be comprehensive and based on integrating our energy use with nature's management of carbon, water, and life on earth as represented by the carbon and hydrogeological cycles. Obviously, the challenges presented by the need to control atmospheric greenhouse gases are enormous and require "out-of-the-box" thinking, an innovative approach, imagination, and bold engineering initiatives to achieve sustainable development. We will need to ingeniously exploit even more energy and integrate its use with control of atmospheric greenhouse gases. The continued development and application of energy is essential to the development of human society in a sustainable manner through the coming centuries. All alternative energy technologies are not equal and have risks and drawbacks. When evaluating our energy options, we must consider all aspects, including performance against known criteria; basic economics and benefits; efficiency; processing and utilization requirements;

infrastructure requirements; subsidies and credits; waste and ecosystem; as well as unintended consequences, such as impacts to natural resources and the environment. In addition, we must include the overall changes and the emerging energy picture based on current and future efforts to modify fossil fuels and evaluate the energy return for the investment of funds and other natural resources such as water.

A significant motivation in creating this book series, which is focused on alternative energy and the environment, was brought about as a consequence of lecturing around the country and in the classroom on the subject of energy, environment, and natural resources such as water. Water is a precious commodity in the West in general and the Southwest in particular and has a significant impact on energy production, including alternative sources, due to the nexus between energy and water and the major correlation with the environment and sustainability-related issues. While the correlation between these elements, how they relate to each other, and the impact of one on the other are understood, it is not significantly debated on when it comes to integration and utilization of alternative energy resources into the energy matrix. Furthermore, as renewable technology implementation grows by various states, nationally and internationally, the need for informed and trained human resources continues to be a significant driver in future employment, resulting in universities, community colleges, and trade schools offering minors, certificate programs, and even in some cases majors in renewable energy and sustainability. As the field grows, the demand for trained operators, engineers, designers, and architects who would be able to incorporate these technologies into their daily activity is increasing. Also, we receive a daily deluge of flyers, e-mails, and texts on various short courses available for interested parties in solar, wind, geothermal, biomass, and so on under the umbrella of retooling an individual's career and providing trained resources needed to interact with financial, governmental, and industrial organizations.

In all my interactions throughout the years in this field, I have conducted significant searches in locating integrated textbooks that explain alternative energy resources in a suitable manner and that would complement a syllabus for a potential course to be taught at the university while providing good reference material for people interested in this field. I have been able to locate a number of books on the subject matter related to energy; energy systems; resources such as fossil, nuclear, renewable, and energy conversion; as well as specific books in the subjects of natural resource availability, use, and impact as related to energy and the environment. However, specific books that are correlated and present the various subjects in detail are few and far between. We have therefore started a series of texts, each addressing specific technology fields in the renewable energy arena. As a part of this series, there are textbooks on wind, solar, geothermal, biomass, hydro, and other subjects yet to be developed. Our texts are intended for upper-level undergraduate and graduate students and for informed readers who have a solid fundamental understanding of science and mathematics as well as individuals/organizations involved with design development of the renewable energy field entities that are interested in having reference material available to their scientists and engineers, consulting organizations, and reference libraries. Each book presents fundamentals as well as a series of numerical and conceptual problems designed to stimulate creative thinking and problem solving.

I wish to express my deep gratitude to my wife, Maryam, who has served as a motivator and intellectual companion and too often was victim of this effort. Her support, encouragement, patience, and involvement have been essential to the completion of this series.

Abbas Ghassemi, PhD
Las Cruces, New Mexico

Preface

Solar radiation transfer through the Earth's atmosphere and interaction of solar radiation with the land, air, and oceans are recognized as the drivers for the Earth's radiation budget. As a result, there is a large, active climate change modeling community devoted to highly detailed, complex, and often esoteric physical processes relating to solar radiation interactions with the atmosphere. The immense energy resource potential presented by the solar radiation reaching the Earth's surface has been long recognized. The highly visible products of the climate change scientific community are far too detailed and complex for the renewable energy enthusiast, student, engineer, designer, or financier to acquire, implement, or understand.

Given the importance of understanding the "fuel resources" for renewable, and especially solar, energy conversion system, it is surprising how little attention has been paid over the past 50 years to the practical modeling of solar radiation resources for renewable energy applications. The number of scientists and engineers devoted to the field of modeling terrestrial solar radiation for renewable energy applications has been relatively small, but productive. At best, the methods, models, and issues associated with this topic are widely distributed through the scientific literature and specialized reports. Many specialized topics are covered only in obscure technical reports, solar conference proceeding papers from a small number of experts participating in short or limited sessions. This literature is not easily accessible, especially by the student or the general public. There have been attempts at developing regular periodic conferences, and specialized scientific journals devoted solely to solar radiation for renewable applications have been sporadic, but to date, those attempts have been unsuccessful.

The purpose of this book is to bring the nuts and bolts of practical solar radiation modeling applications to the student, professional working scientist or engineer, and solar energy conversion system enthusiast. The organization and content are based on 30 years of active participation in the solar energy measurement and modeling field. The intent is to answer many of the frequently asked "how to" questions heard over the years and recently, as interest in solar radiation for many different applications has increased. Primarily, these questions relate to solar energy conversion systems, solar day lighting, and energy efficiency of buildings. However, there is also a wide variety of other scientific, practical, and even artistic applications. The alternative fields most often used include human health and vision, optical properties of materials, agriculture, influences on plants and animals, and even global climate change. The "answers" provided here are mostly "numerical recipes" that the reader can implement personally or find on the World Wide Web and use both to learn from and to meet his or her solar energy conversion-related modeling needs.

The structure of the book is designed to introduce the fundamentals of solar radiation resource estimation. The solar radiation at the top of the atmosphere, the impact of the ever-changing but predictable path of the sun through the dome of the sky, and the highly variable atmospheric filter are described. A brief discussion of solar radiometers, their calibration, and accuracy of their measurements is presented. We then move to the heart of the book, the modeling of solar radiation at the ground. We begin with the easier case of solar radiation under clear skies. Then, the modeling of solar radiation under partly cloudy conditions is addressed. Next, the modeling of solar radiation available for various solar energy collector geometries and the modeling of missing solar radiation data elements when only a few pieces of data or information are available are described. A brief overview of the simplest models for estimating the distribution of solar energy as a function of wavelength of light is given. The book concludes with chapters on the modeling of spatial distribution of daylight over the sky dome for architectural lighting computations.

I am grateful for the opportunity afforded over the past 35 years to learn from, associate with, and interact with the small number of dedicated scientists devoted to solar radiation modeling for renewable energy applications. These ladies and gentlemen have become friends as well as professional colleagues. Any errors, omissions, or mistakes in general are my sole responsibility.

Daryl Ronald Myers
August 2012

Acknowledgments

Most important, my loving and patient family and spouse have long endured my devotion, sometimes to the point of distraction, to my career. It has been my personal and professional interaction with many colleagues over many years that inspired this book. It is impossible to convey the importance of the collegial and truly friendly interaction I have personally enjoyed with the solar resource assessment and optical radiation physics community. These colleagues actively contributed to my interest and education in the field of solar radiation measurements and resource assessment. Many, regrettably, have slipped the mortal coil and now leave only the legacy of their work. Funding for my entire career (1974 to 2011) has been provided by the U.S. Department of Energy and the Smithsonian Institution and therefore by the taxpayers of the United States. With this support, it has been a privilege to work with and for colleagues from the National Renewable Energy Laboratory (NREL), including, in no particular order, Roland L. Hulstrom, Chester V. Wells, Dr. Eugene Maxwell, Dr. Martin Rymes, Dr. David Renné, Thomas L. Stoffel, Stephen Wilcox, Ibrahim Reda, Dr. Carol Riordan, Dr. Theodore Cannon, Keith Emery, Carl Osterwald, William Marion, Dr. Sarah Kurtz, Dr. Jerry Olson, and Joseph Del Cueto. Among those in the wider solar modeling and measurement government, academic, and industrial community, I owe great debt to John R. Hickey of the Eppley Laboratory; Gene Zerlaut, formerly of Desert Sunshine Exposure and Test Laboratories; Joe Robbins of Arizona Desert Testing LLC; Dr. Frank Vignola of the University of Oregon; Dr. Christian Gueymard of Solar Consulting Services; Dr. Richard Perez of the State University of New York, Albany; Stephen Hogan of Spire Corporation; Dr. Richard Young of Optronics Laboratories; Dr. David Mennicucci of Sandia National Laboratories; Dr. Yoshi Ohno of the National Institute of Standards and Technology (NIST); Dr. Ellsworth Dutton and Dr. Joseph J. Michalsky, National Oceanic and Atmospheric Administration (NOAA); and Dr. Paul Stackhouse and Dr. Thomas P. Charlock, National Aeronautics and Space Administration (NASA); all have provided invaluable insight and enlightenment.

The international quest to improve solar measurements and modeling brought new perspectives on the field by personal discussions and exchanges with Dr. Klaus Dehne, German Weather Service; Dr. Claus Fröhlich, Physical Meteorological Observatory Davos, Switzerland; Dr. John Page, Emeritus Professor, University of Sheffield, United Kingdom; Dr. Saleh Al-Awaji, Dr. Naif Al-Abaddi, and Mohammed Bin Mahfood, King Abdul Aziz City for Science and Technology, Riyadh, Kingdom of Saudi Arabia; Dr. Harry Kambezidis, National Observatory of Athens, Athens, Greece; Dr. Tariq Muneer, Napier University, Edinburgh, Scotland; Dr. Wang Bingzhong, Chinese Academy of Meteorological Services, Beijing, People's Republic of China; and Dr. Anatoly Tsvetkov, Voeikov Main Geophysical Observatory, St. Petersburg, Russian Federation.

Among those no longer with us—Dr. Richard Bird of NREL; Edwin C. Flowers of NOAA; Warren Ketola of 3M; Dr. William H. Klein, Bernard Goldberg, and

Douglass Hayes of the Smithsonian Radiation Biology Laboratory; and Dr. William Saunders of NIST—all provided inspiration, education, and friendly advice over many years of fruitful collaboration. Finally, thank you to Ashley Gasque of Taylor & Francis, CRC Press, for asking the question.

About the Author

Daryl R. Myers is a native Coloradoan. He has a BS in applied mathematics from the University of Colorado School of Engineering. During the Cold War, he served 4 years in the U.S. Army as a Russian linguist. He spent a further 4 years as a physicist at the Smithsonian Institution Radiation Biology Laboratory. In 1978, he joined the Solar Energy Research Institute, now the National Renewable Energy Laboratory (NREL). Daryl contributed to development of many national solar radiometry consensus standards, the U.S. National Solar Radiation Data Base, and joint satellite data validation projects in conjunction with the National Aeronautics and Space Administration (NASA) Earth Observing Systems and the King Abdul Aziz City for Science and Technology (KACST) in the Kingdom of Saudi Arabia. After 37 years of work in solar radiation research, he retired from NREL in 2011. He is author or coauthor of 160 technical publications. He is a former member of the American Society for Testing and Materials, the International Lighting Commission Division 2 on Physical Measurement of Light and Radiation, a former president of the Council for Optical Radiation Measurements, and editor of the council's newsletter *Optical Radiation News*. His primary recreational interests are fly-fishing in the high country of Colorado, amateur astronomy and astrophysics, military history, and the history of science and technology, particularly naval, aeronautical, and space technology. He lives in a suburb of Denver, Colorado, with his wife, Barbara.

1 Fundamentals of Solar Radiation

> The Sun remains fixed in the center of the circle of heavenly bodies, without changing its place; and the Earth, turning upon itself, moves round the Sun.
>
> **—Galileo Galilei, 1615**

1.1 THE SUN AS A STAR

Earth orbits a star, the sun, which is the ultimate source of all energy driving the process of animate and inanimate life cycles on the planet. The sun's nuclear furnace continually fills the volume of surrounding space with energetic elementary particles and photons of electromagnetic radiation. The sun's electromagnetic spectrum spans an enormous range of wavelengths of frequencies of radiation, from gamma and x-rays, to ultraviolet (UV), visible, infrared (IR), and radio waves. For the purposes of this book, we are interested in so-called optical solar radiation, from the UV wavelengths to the near- and mid-infrared wavelengths that Earth's atmosphere allows to reach the ground. We denote this region of interest as optical solar radiation even though only a subset of the spectrum, that within the photopic response of the human eye, is "optically visible." The details of the distribution of optical radiation as a function of wavelength are discussed in this chapter.

1.2 THE EARTH AND THE SUN

1.2.1 THE ORBIT AND ROTATION OF THE EARTH

Earth orbits the sun in a very slightly elliptical orbit, with an eccentricity ε (ratio of major to minor axis) of 0.0167. Earth also rotates once every 23 h 56 min, with respect to the distant stars (a sidereal day), and a period of 24.0 h (the definition of the solar day) about an axis titled at an angle of 23.5° to the plane of that orbit. The average distance between the Earth and the sun is the astronomical unit (AU) of 93 million miles (or 149,597,870.7 km).

The tilted axis of rotation produces the seasonal weather changes we experience, as shown in Figure 1.1. That tilt also causes the daily changes in the points on the horizon where the sun rises and sets, the path of the sun through the sky dome, and the period of daylight to change throughout the year. The eccentricity of the orbit produces changes in the Earth–Sun distance, or "radius vector" r, with respect to the mean distance of 1 AU. The closest approach of the Earth to the sun (perihelion =

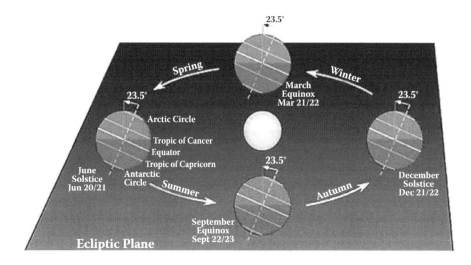

FIGURE 1.1   Earth's orbit and the seasons. (From Wikipedia, http://upload.wikimedia.org/wikipedia/commons/thumb/f/f0/Seasons1.svg/700px-Seasons1.svg.png.)

147.09×10^6 km) is around the first week in January, and the greatest Earth–Sun distance (152.1×10^6 km) is around the first week of July. The actual dates vary from year to year, depending on leap years, and small cumulative geophysical or gravitational influences [1].

1.2.2 THE SUN AND INTENSITY OF EXTRATERRESTRIAL SOLAR RADIATION

Earth intercepts only a minute fraction of the energy radiated by the sun into the surrounding spherical volume of space. The "flux density" or "intensity" of the radiation, in terms of watts per square meter of area (Wm^2) at distance R from the sun falls off as $1/R^2$. Investigation of the intensity Io of the solar radiation at the average distance of the Earth's orbit, or so-called extraterrestrial radiation (ETR), has long been a subject of scientific investigation. During the space age, satellite-based estimates of the intensity of solar radiation at 1 AU vary around Io = 1366 Wm^2 with an uncertainty of about ±7 Wm^2. For most of the period from 1970 to 2010, the accepted value of Io was 1366.1 Wm^2 [2,3].

In 2011, instrumentation issues with historical measurements in conjunction with new measurements from space produced a value of Io = 1361 Wm^2 [4]. Note that this difference amounts to ±0.5%, and the choice of Io in the model scenarios that follow change depending on which value of Io the user selects.

The eccentric Earth orbit results in a +3% increase in the solar radiation intensity at perihelion and a –3% decrease at aphelion. This variation can be accounted for by relying on detailed astronomical calculations such as appear in the annual astronomical or nautical almanacs published for astronomers and navigators relying on classical techniques [5,6]. Two popular equations for calculating the Earth–Sun radius vector correction term Rc are [7,8]

$$Rc = (Ro/R)^2 = 1.000110 + 0.034221\ \cos(d) + 0.001280\ \sin(d)$$
$$+ 0.000719\ \cos(2d) + 0.000077\ \sin(2d) \tag{1.1}$$

where Ro is the mean distance, R is the actual distance, and d is the "day angle" computed from

$$d = 2\pi(dn - 1)/365, \text{ with } dn = \text{day of the year (Jan 1 = 1)}$$

and

$$Rc = 1 + 0.033\ \cos(2\pi dn/365) \tag{1.2}$$

The solar constant Io is modified by Rc to produce the ETR for each day of the year as ETR = Io Rc.

1.3 SOLAR TIME AND SOLAR POSITION

The computation of the "solar position" or location of the sun in the hemisphere of the sky and the geometrical relationship between the sun and the Earth's surface or planes and surfaces for collecting solar energy is a subject covered by many authors. The methods derived and reported can vary widely in accuracy, with the benchmark or reference standard usually being the computations used to produce the astronomical almanacs mentioned. There have been implementations of the astronomical almanac algorithms developed and implemented online, such as the Solar Position and Intensity (SOLPOS) calculator of Michalsky [9] or the solar position algorithm (SPA) developed by Reda and Andreas [10,11]. While SPA is very accurate, this algorithm is too extensively complex to include here but is well worth studying.

Given that the diameter of the solar disk subtends an angle of 0.52° (aphelion) to 0.54° (perihelion) in the sky, accuracy of 0.1° to 0.25° in pointing to or tracking the sun in the sky for solar energy applications is sufficient. Given this tolerance, there are several popular simple, acceptable equations for computing the solar position.

The solar position is usually denoted by solar elevation e, or angular distance above the horizon, and the solar azimuth α, or angle in the horizontal plane of the observer from a reference direction, either north = 0 or south = 0, to the center of the solar disk as seen from a location on Earth.

1.3.1 Solar Time

Solar time is based on the apparent position of the sun in azimuth with respect to the local meridian, or line of longitude, Ψ. Standard time was developed to partition Earth into 24 "time zones," each approximately 15° in width. The reference or starting point for standard time is the line of zero longitude running through Greenwich, England, which is the "standard meridian" for time zone 0. The standard meridian Ψs for each other time zone runs through the center (±7.5° from the boundaries) of the zone. Zones are numbered positive for those east of Greenwich and negative for those west of Greenwich.

It is important to note that for civil, geopolitical, or practical reasons, designated boundaries of time zones deviate significantly from the ±7.5° theoretical lines of longitude for the boundaries. For some locations, time zone designators are also in terms of ½-h intervals. Thus, quoted "civil time zone" designations for sites near zone boundaries or in special "1/2 time zones" may need adjustment by plus or minus 1.0 or 0.5 time zone for correct solar geometry calculations. It is always the true longitude Ψ of the site, with respect to the Greenwich meridian $\Psi = 0°$, and the standard meridian Ψs of the theoretical integral time zone number for the site that should be used to designate the time zone in solar position calculations.

1.3.2 DECLINATION

Because of the 23.5° axial tilt of Earth's rotation axis (with respect to the plane of Earth's orbit), the location (azimuth from north) of sunrise and sunset on the horizon, the path of the sun, and the length of the solar path from sunrise to sunset (or day length in hours) vary for each location on a daily basis throughout the year (see Figure 1.1).

The projection of Earth's equator onto the sky dome is the equatorial plane. The angle between the equatorial plane and the plane of Earth's orbit (or the apparent orbit of the Sun about the Earth or the "plane of the ecliptic") changes as the Earth moves along its orbit through the year. This angle is the declination δ of the sun (e.g., "deviation" between the projected equatorial plane and the orbital plane).

At the equinoxes (spring and fall, equal day length of 12 h, the Sun at the intersection of the equatorial and ecliptic planes), the declination is zero. At the solstices (winter and summer, shortest and longest day length, respectively), the Sun is at the greatest declination or deviation below (–23.5°) and above (+23.5°) the equatorial plane. The declination of the sun can be computed from [7]

$$\delta = 0.006918 - 0.399912 \cos(d) + 0.070257 \sin(d) - 0.006758 \cos(2d)$$
$$+ 0.000907 \cos(2d) - 0.002697 \cos(3d) + 0.00148 \sin(3d) \tag{1.3}$$

where δ is in radians (multiply by 57.296 to get degrees), and d is the same day angle used for Equation 1.1.

A simpler formula for δ, sufficient for engineering calculations, is [8]

$$\delta = 23.45° \sin([360/365][dn + 284]) \tag{1.4}$$

where dn is the day number, 1 to 365.

1.3.3 EQUATION OF TIME

The time it takes Earth to traverse equal distances along its elliptical orbit varies through the year, while the daily rotation rate is constant. Thus, the local standard time (progressing at a uniform rate) that the sun is located on the local meridian (solar noon) will vary through the year. The difference between standard time and solar time is called the equation of time E_t. The equation of time varies from the sun being 14 minutes "slow" (crossing the local meridian after noon standard time) in January to 16 minutes "fast" (crossing the local meridian before noon standard time) in October. E_t can be computed from [7]

$$E_t = 0.0000075 + 0.001868 \cos(d) - 0.032077 \sin(d)$$
$$- 0.014615 \cos(2d) - 0.040849 \sin(2d) \qquad (1.5)$$

In this case, E_t is actually computed in terms of angular units of radians. To convert to units of minutes of time E_{tm}

$$E_{tm} = E_t(229.18) \qquad (1.6)$$

1.3.4 LOCAL APPARENT TIME

The conversion between solar time, or "local apparent time" (LAT) and local standard time is a function of the difference between the longitude of the location Ψ, the longitude of the time zone standard or central meridian Ψs, and the equation of time [12].

$$\text{LAT} = \text{Local standard time (LST)} + E_{tm} + 4(\Psi s - \Psi) \qquad (1.7)$$

1.3.5 SOLAR POSITION: ALTITUDE AND AZIMUTH

The parameters that define the solar position are the solar elevation e and the solar azimuth α. In solar energy applications, the complement of the solar elevation, $90° - e$, called the zenith angle z, is often convenient. The position of the sun in the sky at a given standard time is a function of the latitude φ; longitude ψ (in terms of the difference in longitude between the longitude of the location and the standard meridian for the time zone of the site); the equation of time E_t; and the declination δ. These parameters are all functions of the day of the year. Figure 1.2 shows the apparent solar path for a midlatitude Northern Hemisphere site for the months from June to December. The paths for January to May vary slightly from those shown here. The paths in the figure were calculated using the algorithms provided in the following discussion in terms of solar elevation angles and solar azimuth angles.

The LAT must be expressed as an angle, called the hour angle ω, of the sun, defined (for solar applications) as ω = zero at solar noon and positive or negative before noon and after noon, respectively. The hour angle is $15°$ (= $360°/24$-h day length) multiplied by the (fractional) hours away from solar noon [12]:

$$\omega = 15°(\text{Local apparent noon} - \text{LAT}) = 15°(12.0 - \text{LAT}) \qquad (1.8)$$

By applying the law of cosines for spherical triangles, it can be shown that the sine of the solar elevation angle e or cosine of the solar zenith angle z (angle between the normal to a local horizontal and the direction of the sun in the sky) is given by [12]

$$\sin(e) = \cos(z) = \sin(\varphi) \sin(\delta) + \cos(\varphi)\cos(\delta)\cos(\omega) \qquad (1.9)$$

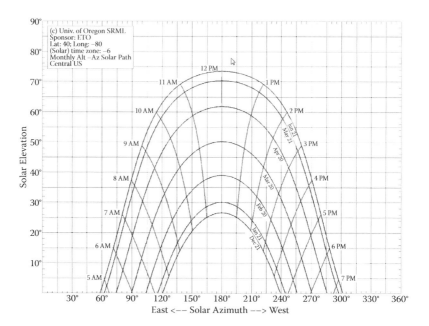

FIGURE 1.2 Sun path for 40 N 80 W in the central United States in altitude-azimuth format. Note the typical geographical convention of south = 180° azimuth. (From University of Oregon, http://solardat.uoregon.edu/SunChartProgram.php, accessed 16 June 2012.)

To find the solar azimuth angle α, compute the sine of the auxiliary angle α_s:

$$\sin(\alpha_s) = \cos(\delta)\sin(\omega)/\cos(e) \tag{1.10}$$

The angles e, z, and α_s are the inverse trigonometric functions of the results of Equations 1.9 and 1.10.

Since α is 180° when the sun is due south (hour angle = 0), less than 180° in the morning, and greater than 180° in the afternoon, and $\alpha_s = 0$ when the hour angle ω is zero,

$$\alpha = 180 - \alpha_s. \tag{1.11}$$

Note when the sun is at an elevation of 90°, Equation 1.10 fails. If the sun is directly overhead, the azimuth angle is undefined. The prudent programmer should check to see if the denominator approaches zero to prevent divide-by-zero errors.

1.3.6 Solar Incidence Angles

Solar radiation propagating through the atmosphere is attenuated by absorption and scattering. Various molecules in the atmosphere, in particular water vapor and suspended or airborne particles, influence these processes. These atmospheric elements decompose the solar radiation into constituents referred to as "solar radiation components," discussed in the next section. The relative magnitude of these components is a function of the path length for the beam radiation through the atmosphere. In turn, this path length is a function of the position of the sun in the sky and geometrical relationships between the plane of interest and the sun. For a horizontal surface, the incidence angle of the direct beam is the complement (90 − e) of the elevation angle, obtained from Equation 1.9. This is the zenith angle z, which is the angle between the normal to a horizontal surface and the line from the foot of the normal to the solar disk.

For a tilted surface, the incidence angle computation becomes more complex. The tilt angle (angle between horizontal and the plane of the surface), azimuth angle of the plane (angle between the projection of the normal to the surface on a horizontal plane and either north = 0° or south = 0°, depending on the convention selected), and the azimuth angle of the solar disk must be accounted for.

The general equation for the cosine of the incidence angle θ of the direct beam on a surface tilted at slope angle s and with the normal to the surface at azimuth γ *with respect to south = 0 azimuth* is

$$\cos(\theta) = \sin(\delta)\sin(\varphi)\cos(s) - \sin(\delta)\cos(\varphi)\sin(s)\cos(\gamma) + \cos(\delta)\cos(\varphi)\cos(s)\cos(\omega)$$
$$+ \cos(\delta)\sin(\varphi)\sin(s)\cos(\gamma)\cos(\omega) + \cos(\delta)\sin(s)\sin(\gamma)\sin(\omega) \quad (1.12)$$

Note that if the azimuth angle of the surface normal γ is 0, or due south, this reduces to

$$\cos(\theta) = \sin(\delta)\sin(\varphi)\cos(s) - \sin(\delta)\cos(\varphi)\sin(s) + \cos(\delta)\cos(\varphi)\cos(s)\cos(\omega)$$
$$+ \cos(\delta)\sin(\varphi)\sin(s)\cos(\omega) \quad (1.13)$$

For the horizontal case, tilt angle s = 0, and azimuth angle 0

$$\cos(z) = \cos(\theta) = \sin(e) = \sin(\delta)\sin(\varphi) + \cos(\delta)\cos(\varphi)\cos(\omega) \quad (1.14)$$

In all cases, the incident angle per se is the inverse trigonometric function of the results of Equation 1.12, 1.13, or 1.14.

1.3.7 Air Mass

As mentioned (Section 1.3.6), the relative magnitudes of the solar components are functions of the path length through the atmosphere. If the atmosphere is treated as a horizontal slab of material of unit thickness, the shortest path length is along the vertical to the local horizontal. This path length, relative to zenith angle z = 0 or solar elevation e = 90, is defined as "air mass," M. This term is not to be confused with the meteorological air mass, which represents high- and low-pressure atmospheric pressure areas or storms.

Geometrically, as the zenith angle increases, the path length relative to the vertical increases as 1/cos(z) or 1/sin(e). Thus,

$$M = 1/\cos(z) = 1/\sin(e) \quad (1.15)$$

For example, for z = 60, M = 1/cos(60) = 1/0.5 = 2.0, as shown in Figure 1.3.

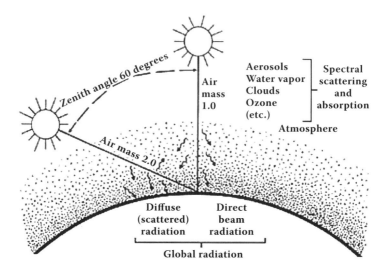

FIGURE 1.3   Air mass, atmospheric constituents, and radiation components. (From NREL, http://rredc.nrel.gov/solar/pubs/shining/chap3_sub.html.)

As discussed in the following material, many components of solar radiation models are parameterized with respect to air mass. The actual physical thickness of the atmosphere above a site is a function of the altitude of the site. The geometrical air mass is with respect to an elevation of sea level, with an atmospheric pressure Po of 101.324 hectapascals (hPa) or 1013.25 millibar (mb). The local atmospheric pressure or station pressure Ps is directly related to the site altitude and the thickness of the atmosphere above the station. In the absence of measured station pressure data, a good approximation of Ps as a function of altitude h in kilometers is

$$Ps = Po\ e^{-0.000832\,h} \tag{1.16}$$

The pressure-corrected air mass Mp, representing the actual physical path length relative to z = 0 and sea level, is

$$Mp = MPs/Po \tag{1.17}$$

In reality, Earth's atmosphere is a spherical shell surrounding the planet. Also, the index of refraction for space (= 1.0) and the atmosphere (~1.0002772) is different. This means that just as a rod inserted into a pool of water appears bent at the air–water interface (due to the different indices of refraction of water and air), the ETR beam radiation is bent at the air–space interface. This bending is the refraction of the beam by the atmosphere. Astronomers are quite familiar with this effect, which must be accounted for when pointing telescopes at celestial objects seen through the atmosphere at high zenith angles.

Since the radius of curvature of the spherical atmosphere is so large, the horizontal slab approximation for the atmosphere is reasonable for zenith angles less than about 70°. At larger zenith angles, near sunrise or sunset, the effect of refraction is larger. At sunrise or sunset, when the bottom of the solar disk is apparently tangent to the horizon, if there were no atmosphere, the top of the solar disk would be tangent to the horizon. Thus, the solar disk is entirely below the horizon at a geometrical zenith angle of 89.5°. The refracted or true zenith angle at such times is actually 90.25°. This means the true, refracted air mass, or path length for the photons at large zenith angles Mr, is larger than geometrical air mass M calculated solely from geometry, as in Equation 1.15 or 1.17.

There have been many publications addressing the computation of true zenith angles and air mass at high zenith angles. The most widely accepted correction algorithm for computing Mr from the apparent zenith angle z is that of Kasten and Young [13]:

$$Mr = 1/(\cos(z)) + 0.50572*(96.07995 - z)^{-1.6364} \tag{1.18}$$

The refraction-corrected air mass, MrP, is then corrected for the site station pressure:

$$Mrp = MrPs/Po \tag{1.19}$$

1.4 SOLAR COMPONENTS

The portion of the ETR arriving at Earth's surface is the direct radiation, also called direct beam radiation B or direct normal irradiance (DNI). DNI is radiation on a plane normal or perpendicular to the line connecting the observer and the center of the solar disk, within a small solid angle, usually 5° or smaller, which includes the solar disk.

Diffuse radiation (D) is solar radiation scattered out of the direct beam by the atmosphere into the hemisphere of the sky dome. The diffuse irradiance on a horizontal plane we denote as DHI. The total ("global") hemispherical radiation on a plane, G or GHI, is the combination of the direct normal radiation multiplied by the cosine of the incidence angle θ (between the normal to the plane and the direction from the base of the normal to the center of the solar disk), DNI Cos(θ), plus the diffuse sky radiation:

$$GHI = DNI \cos(\theta) + DHI \tag{1.20}$$

or

$$G = B \cos(\theta) + D \tag{1.21}$$

If a flat receiver collector plane is tilted away from the horizontal, reflected ground radiation Rg must be added to the diffuse sky radiation seen by the tilted plane, DTI, to obtain the global (total) hemispherical irradiance on the tilted plane, GTI (see Figure 1.4):

$$GTI = DNI \cos(\theta) + DTI + Rg \tag{1.22}$$

DNI, DHI, GHI (or B, G, D), and Rg are the basic solar radiation "components."

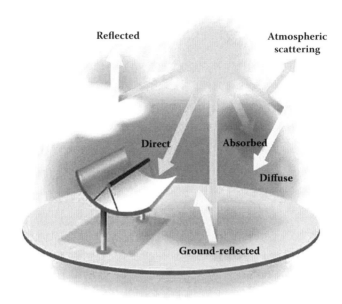

FIGURE 1.4   Solar components. (From NREL, http://www.nrel.gov/docs/fy10osti/47465. pdf (page 7).)

For a plane tracking the sun, the total hemispherical irradiance (global normal irradiance) is GNI. GNI is the sum of the direct beam B, ground reflected irradiance Rg intercepted by the plane, and the diffuse sky radiation DTI intercepted by the tracking (tilted) plane.

$$GNI = DNI + DTI + Rg \tag{1.23}$$

Typical solar models for estimating DTI and Rg components are described in Chapter 6.

1.5   CLEARNESS INDEX

We now have defined most of the parameters used in modeling propagation of solar radiation from the top of the atmosphere to the ground and intercepted by various collector surfaces. Some additional parameters specific to individual models are addressed as needed. However, there is one other useful concept utilized in many solar radiation models. This is the idea of the clearness index [14]. The clearness index is an indication of the bulk transmittance of the atmosphere, even one containing clouds.

Given the solar constant of 1366.1 Wm^2 and the variation in this value as a function of the day of the year, described in Section 1.2.2, the extraterrestrial beam irradiance Io is easily calculated. For a given location, date, and time, the zenith and incidence angles for various collector geometries, as described in Section 1.3.6, may be calculated. The cosine of these incidence angles, applied to the ETR DNI, represents the maximum possible direct beam component (usually the largest component) incident on horizontal or tilted surfaces.

The clearness index for a particular solar component on the ground is computed as the ratio of the measured (or modeled) value to the "equivalent" maximal component based on the ETR above the atmosphere. Thus, for horizontal surfaces

Kt = "Total" global hemispherical component on horizontal surface clearness index
   = GHI/(Io cos(z))   (1.24)

Kd = Diffuse hemispherical component on horizontal surface clearness index
   = DHI/(Io cos(z))   (1.25)

Kn = DNI clearness index = DNI/Io   (1.26)

From the fundamental relation between solar components (Equations 1.20 and 1.21), we see that there is a simple relationship between the direct beam and horizontal surface clearness indices, namely:

$$Kt = Kn + Kd \qquad (1.27)$$

We will see that the clearness indices can be useful parameters in modeling missing solar components from available measured data. Sometimes, models may use the clearness index with respect to tilted surfaces to develop models for solar radiation available to specific solar collector configurations.

1.6 SUMMARY

This chapter introduced the basic solar radiation components:

Io   solar constant of 1366.1 Wm^2
ETR extraterrestrial (beam) solar radiation at the top of the atmosphere, Io
B   direct normal irradiance (DNI) or direct beam irradiance
D   diffuse sky irradiance
DHI total diffuse hemispherical irradiance on a horizontal surface
DTI total diffuse hemispherical irradiance on tilted surface
G   total hemispherical irradiance
GHI total (global) hemispherical irradiance on a horizontal surface
GTI total (global) hemispherical irradiance on a tilted surface
GNI total (global) hemispherical irradiance on a sun-tracking (normal incidence) plane
h   site elevation in kilometers
Kt   total hemispherical clearness index for horizontal surface
Kn   direct beam clearness index
Kd   diffuse hemispherical clearness index on a horizontal surface
Rg   ground-reflected irradiance

The basic parameters needed to compute solar radiation at a specific time and place have been described:

Rc   Earth–Sun distance (radius vector) correction term
dn   day of the year (Jan 1 = 1)
d   "day angle" computed from $2\pi(dn - 1)/365$
φ latitude
Ψ longitude
Ψ_s time zone (Tz) central meridian
δ solar declination
E_t equation of time
LAT   local apparent (solar) time
ω solar hour angle
e   solar elevation angle
α solar azimuth angle
z   solar zenith angle
θ incidence angle for the DNI or beam radiation
M   (geometrical) air mass at sea level
Mp   stationary pressure-corrected air mass
Mr   refraction-corrected air mass
Mrp   refraction- and station pressure-corrected true air mass.

The Lambert/Beer transmittance law in an absorbing medium at position L within the medium (with respect to the entrance surface, L = 0) is $I = Io\ e^{-xL}$ where L is the path length through the medium, Io is the incident irradiance, x is the absorption coefficient (absorption per unit length) of the medium, and I is the irradiance at position L within the medium [15,16]. This law is also the basis for the transmittance of Earth's atmosphere. From our list of parameters, it should be clear that if one computes Io, Rc, θ, and Mrp for a given location, date, and time, the DNI or B at the surface is of the form

$$B = Io\ Rc\ e^{-x\,Mrp} \tag{1.28}$$

where x is an absorption coefficient or transmittance function per unit (pressure-corrected) air mass, or path length Mrp. Establishing an estimate of the magnitude or function x is the second step in modeling direct beam solar radiation at the earth's surface. The next steps involve finding the diffuse sky component and perhaps ground-reflected components of the solar radiation and combining them with the transmitted beam radiation.

The steps mentioned apply only to clear or cloudless sky conditions, where much less-complicated processes occur. Modeling solar radiation under general sky conditions and for various solar collector geometries involves much more complicated processes. Chapter 3 and those following describe in detail some of the most popular methods of accomplishing these processes. But first, there is a brief side trip into the world of solar radiation measurements and their uncertainties.

REFERENCES

1. Cox, A.N. (ed.). (2000). *Allen's Astrophysical Quantities*, 4th ed. AIP Press, Springer, New York.

2. Hickey, J.R., B.M. Alton, H.L. Kyle, and D. Hoyt. (1988). Total solar irradiance measurements by ERB/Nimbus 7. A review of nine years. *Space Science Reviews*, Vol. 48, No. 3–4, pp. 321–342.

3. Fröhlich, C., and J. Lean. (2004). Solar radiative output and its variability: Evidence and mechanisms. *The Astronomy and Astrophysics Review*, Vol. 12, pp. 273–320.

4. Kopp, G., and J. Lean. (2011). A new, lower value of total solar irradiance: Evidence and climate significance. *Geophysical Research Letters*, Vol. 38, L01706, 7 pp.

5. *Nautical Almanac*. United Kingdom Hydrographic Office, Her Majesty's Nautical Almanac Office, Rutherford Appleton Laboratory, under the supervision of S.A. Bell and C.Y. Hohenker. (http://astro.ukho.gov.uk/nao/publicat/na.html)

6. *Astronomical Almanac*. United States Naval Observatory, U.S. Government Printing Office, under the supervision of S. Urban and R.J. Miller. (http://www.usno.navy.mil/USNO/astronomical-applications/publications/astro-almanac)

7. Spencer, J.W. (1971). Fourier series representation of the position of the sun. *Search*, Vol. 2, No. 5, p. 172.

8. Duffie, J.A., and W.A. Beckman. (1974). *Solar Energy Thermal Processes*. Wiley Interscience, Hoboken, NJ.

9. Michalsky, J.J. (1988). The Astronomical Almanac's algorithm for approximate solar position (1950–2050). *Solar Energy*, Vol. 40, No. 3, pp. 227–235.

10. Reda, I., and A. Andreas. (2008). *Solar Position Algorithm for Solar Radiation Applications*. NREL Technical Report NREL/TP-560–34302. National Renewable Energy Laboratory, Golden, CO. http://www.nrel.gov/docs/fy08osti/34302.pdf. Accessed 21 July 2012.

11. Reda, I., and Andreas, A. (2004). Solar position algorithm for solar radiation applications. *Solar Energy*, Vol. 76, No. 5, pp. 577–589.

12. Iqbal, M. (1983). *An Introduction to Solar Radiation*. Academic Press. Toronto.

13. Kasten, F., and A. T. Young. (1989). Revised optical air mass tables and approximation formula. *Applied Optics*, Vol. 28, No. 22, pp. 4735–4738.

14. Liu, B.Y.H., and R.C. Jordan. (1960). The interrelationship and characteristic distribution of direct, diffuse and total solar radiation. *Solar Energy*, Vol. 4, 1–19.

15. Lambert, J.H. (1760). *Photometria sive de mensura et gradibus luminis, colorum et umbrae* [Photometry, or, On the measure and gradations of light, colors, and shade]. Klett, Augsburg ("Augusta Vindelicorum"), Germany. Available online from the University of Strasbourg, France. http://imgbase-scd-ulp.u-strasbg.fr/displayimage.php?album=53&pos=1. See especially p. 391. Accessed 21 July 2012.

16. Beer, A. (1852). Bestimmung der Absorption des rothen Lichts in farbigen Flüssigkeiten (Determination of the absorption of red light in colored liquids). *Annalen der Physik und Chemie*, Vol. 86, pp. 78–88.

2 Introduction to Solar Radiation Measurements

I belong to those theoreticians who know by direct observation what it means to make a measurement. Methinks it were better if there were more of them.

—Erwin Schrödinger, 1989

2.1 OVERVIEW OF DETECTOR TECHNOLOGY

Radiometry is the measurement of radiation. The conversion of photons of light to some sort of measurable signal is the purpose of radiometric detectors. Historically, there have been several means to accomplish this conversion. These include thermal processes and application of the photoelectric effect for solid-state detectors. Thermal processes include calorimetry and thermoelectric effects. For detailed solar radiation measurement histories and information on solar radiometry, see the work of Vignola, Michalsky, and Stoffel [1] or Coulson [2].

2.2 CALORIMETRY

In calorimetry, absorption of photons produces increased thermal energy and a change in temperature of the material. The temperature change before and after exposure of the detector to the radiation is proportional to the energy absorbed and is generally measured in joules or calories. The method requires an accurate means of measuring temperature and knowledge of the physical and thermal properties of the material, namely, the heat capacity Cp. Water (Cp = 1.0 by definition) is often used as the working material since Cp = 1 cal/°C/g [3].

2.2.1 THERMOELECTRIC DETECTORS

2.2.1.1 Thermopiles

A thermoelectric signal is produced by the Seebeck effect [4,5]. An electric potential (voltage) is produced when a "thermojunction" (called a *thermocouple*) is the result of dissimilar metals in contact with each other and in thermal contact with an absorber. A difference in temperature between the sensing junction thermocouple (sensed as the temperature of the absorber) and an identical thermocouple at a reference temperature is required to generate the thermoelectric voltage. Different dissimilar metal combinations are identified as different thermocouple "types." Each thermocouple type produces a different thermoelectric voltage per degree of temperature difference between the reference and sensing junctions. The relationship between the thermoelectric voltage and temperature is also a function of the temperate difference. Over relatively small temperature ranges (about 100°C or 100 K), the voltage/temperature relationship can be considered approximately linear [6].

A commonly used combination is the type T thermocouple, consisting of junctions of copper and constantan. In the range of temperatures encountered in solar radiometers (−40°C to +50°C) and temperature differences developed by black absorbing sensors (about 5°C), the type T thermocouple produces about 40 microvolts per degree centigrade (μV/°C) [6]. To increase the signal to a magnitude that can be measured accurately and reliably, thermojunctions are connected together in series to build a thermopile. The thermopile voltage is directly proportional to the number of junctions in the thermopile. Thus, for a 40-junction type T thermopile in contact with an absorber 5°C warmer than the reference junctions, the expected output voltage is 40×40 μV/°C $\times$ 5°C = 8000 μV or 8 millivolts (mV). This is a reasonable voltage for measurement by modern electronic equipment. Precision miniature thermopiles such as used in solar radiometers are relatively expensive, and the radiometer requires careful design to incorporate the reference junctions, an efficient thermal absorber, protection from the environment for the sensing and reference junctions, and so on.

2.2.1.2 Resistance Detectors

Another thermoelectric property sometimes used is the resistance temperature detector (RTD). This measurement approach relies on the change in resistance of certain materials as a function of temperature. Platinum is the most common example, thus there are platinum resistance temperature detectors (PRTDs). Platinum has a temperature coefficient of 0.358% per degree centigrade. The resistance of the detector element is determined by measuring the voltage drop across the sensor when a known current is passed through the sensor. This is a straightforward application of Ohm's law: R = V/I, where V = voltage (volts), I = current (amperes), and R = resistance (Ohms). However, this approach has the disadvantage of requiring a stable, accurately known current source and accurate resistance measurement equipment [7].

Another example of thermoelectric detectors, rarely used in solar radiometry, is the thermistor. This detector is similar to the RTD in that the temperature coefficient of resistance (TC) is very large. These detectors are often used for ambient temperature and in laboratory temperature measurement and monitoring equipment. A common thermistor temperature coefficient is in ohms per degree centigrade. In this case, the thermistor TC is not a linear function of the resistance itself but a logarithm of the resistance. A common method of converting the resistance to temperature is the Steinhart–Hart equation [8]:

$$T = 1/(a + b(\ln R) + c(\ln R)^2 + d(\ln R)^3) \qquad (2.1)$$

This measurement approach has the same disadvantages as the RTD, namely, an accurate and stable current source and resistance measurement instrumentation. In addition, the coefficients for Equation 2.1 must be derived individually for each thermistor using regression analysis of calibration temperatures versus resistance measurements. This means the statistical properties of the coefficients and overall scatter about the regression equation contribute to the uncertainty in the temperature estimate. Since the detector produces a temperature (or temperature difference measurement), this is essentially a calorimetric measurement technique.

2.2.2 PHOTOELECTRIC DETECTORS

Einstein won the Nobel Prize in Physics in 1905 for explaining the generation of electric current by materials that were illuminated by photons [9]. Electrons in the atoms of the photoelectric material absorb energy proportional to the frequency, or wavelength, of the photons: $E = h\nu = hc/\lambda$, where h is Planck's constant ($6.62606957 \times 10^{-34}$ joule seconds or joule/hertz), λ is the wavelength of the photon or light (meters), ν is the frequency of the light (Hz), and c is the speed of light in a vacuum (2.999×10^6 km/s).

If the photon energy is great enough, the electrons absorb enough energy to escape their atomic orbits and move to the *conduction band*. That is, the electrons move freely about within the matrix of the material and can produce an electric current in an external circuit. The "gap" between the energy keeping the electrons bound to the atom and the conduction band is called the *band gap* of the material [10]. Photons with energy greater than the band gap (typically measured in electron volts, eV) are absorbed and provide enough energy to produce free electrons in the material. The relation between energy, in terms of electron volts and photon wavelength (or frequency), is

$$E = h\nu \qquad (2.2)$$

The relationship between the material band gap and distribution of the wavelengths of light (spectral distribution) hitting the material means that the material will produce free electrons only for photons with a limited wavelength range. This is called the spectral response region of the detectors. The most common solid-state detector used on solar radiometers is crystalline silicon, with a band gap of 1.1 eV. This means photons of wavelength greater than about 1 micrometer (μm) or 1000 nanometer (nm) cannot produce conduction band electrons. Various other materials, including various forms of silicon, such as "metallurgical-grade," multicrystalline, or amorphous silicon, have widely differing spectral response functions that not discussed here.

Silicon detectors are relatively inexpensive and have several disadvantages that affect their accuracy in solar radiometry. First is the limited spectral response region (about 75% of the total solar spectrum, discussed briefly in Chapter 8). The spectral distribution of sunlight is highly variable, and certain parts of the solar spectrum where the variability is high are not sensed by the silicon detector. Thus, information about the variability in the spectrum outside the spectral response region is not contained in the sensor signal.

Second, the signal the detector produces is an electron current. Third, this signal can be relatively small, a few tens of microamperes (μA) per incident unit of power (Wm²). Measurement of current directly is usually less accurate than measurements of voltage. To get a voltage signal that can be measured more accurately, a stable, low temperature coefficient resistor in parallel with the detector can be used. However, this is another component with quality that has an impact on the quality of the measurement. Last, silicon has a variable temperature coefficient of response, which is a function of the wavelength of light.

In the following sections, the issues associated with thermopile and silicon detector radiometers in radiometers for measuring solar irradiance components are discussed.

2.3 PYRHELIOMETERS: MEASURING DIRECT NORMAL IRRADIANCE

2.3.1 PYRHELIOMETER DESIGN

Measuring the direct normal irradiance (DNI) requires an instrument that collects all the photons in the beam, or solid angle, subtended by the solar disk (approximately 0.5° total solid angle, or 0.008 radians, about 10 milliradians [mR]), and blocking off photons from the rest of the hemisphere. This is accomplished with a "Gershun tube" or long narrow tube with baffles that limit the field of view (FOV) of the detector. The detector is placed at the bottom of the tube. The tube blocks off the sky diffuse photons. When the detector/tube combination is pointed at the sun (detector normal to the direct beam), only the photons in the FOV determined by the geometry of the tube length, baffles, apertures, and detector size are captured by the detector. This device is called a *pyrheliometer*. The basic geometry of the tube, baffles, and detector to determine FOV are shown in Figure 2.1 [11].

A requirement for pyrheliometric measurements of the DNI is that the detector always be normal to the direct beam, which means the pyrheliometer must track the sun along its diurnal path throughout the day. Because the DNI beam is rather slender and small, the tracking accuracy requirements needed to ensure that all the photons in the beam get to the detector are very stringent. To reduce the tracking accuracy requirements (as well as the length of tube required to get the desired FOV), the FOV is generally opened up to a larger solid angle. This permits some tracking inaccuracy without losing photons from the beam.

2.3.2 CIRCUMSOLAR RADIATION

For a pyrheliometer with a larger FOV, part of the sky around the sun is also in the FOV of the detector. This region between the limb (edge) of the solar disk and the edge of the FOV is the circumsolar region and contains circumsolar radiation. At first, this circumsolar radiation may seem to introduce an error into the measurement of the DNI. This is not the case because the circumsolar radiation is radiation scattered out of the beam in random directions and is not collected by the pyrheliometer sensor [12].

The direction a scattered photon follows depends on the complex physics of the air molecules and other (variable) components in the atmosphere [10]. There may be some (usually small) part of the radiation scattered in the forward direction along the DNI beam. Direct measurements of the relative intensity or magnitude of the circumsolar radiation show that within a few arc seconds of the edge of the solar disk,

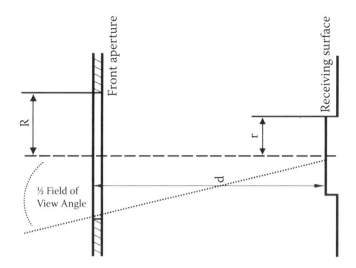

FIGURE 2.1 Field of view defining parameters for pyrheliometers. (Adapted from World Meteorological Organization. 2008. *WMO Guide to Meteorological Instruments and Methods of Observation*, WMO No. 8, 7th ed. World Meterological Organization, Geneva, Switzerland. http://www.wmo.int/pages/prog/gcos/documents/gruanmanuals/CIMO/CIMO_ Guide-7th_Edition-2008.pdf. Accessed 21 July 2012.)

the number of photons (magnitude of the intensity) falls off more than two or three orders of magnitude (factor of 1/100 to 1/1000) below the disk intensity [13,14]. Even if these forward-scattered photons reach the detector, they are "lost in the noise" or random thermal signal from the detector. The requirement for the number of circumsolar photons to exceed the signal-to-noise level is that the atmosphere contain a large number of scattering centers (particles) with sizes on the order of, or greater than, the wavelengths of light present in the solar spectrum. The theory of such scattering is due to Gustav Mie (1908) [15,16]. These conditions result in very large circumsolar regions and indistinct edges of the solar disk.

Discussing a model for extracting the "pure" direct beam intensity from such conditions is beyond the scope of this book. The model developer and user should keep in mind that issues such as the circumsolar contribution may affect the accuracy of the models. Uncertainty in general and for specific instrumentation types is discussed in Section 2.6.

2.4 PYRANOMETERS: MEASURING HEMISPHERICAL RADIATION

Just as the pyrheliometer requires a well-defined FOV, so does the pyranometer, the device used to measure total hemispherical radiation, the so-called global radiation. As the descriptor implies, the FOV of the pyranometer is a hemisphere, 180° from horizon (as seen from the center of the plane of interest) to zenith to horizon and 360° around the horizon, or a 2 π sterradian (sr) solid angle. For horizontal surfaces, the measurement is global horizontal irradiance (GHI). For tilted surfaces, there is global tilted irradiance (GTI).

FIGURE 2.2 (See color insert.)   Thermal (top right) and solid-state photoelectric detector pyranometers including a silicon photovoltaic reference cell (bottom left).

2.4.1 THERMAL PYRANOMETERS

The detector for the pyranometer is thus a flat plane absorber in thermal contact with the sensor. The detector is usually protected by one or more hemispherical domes of glass or quartz (Figure 2.2). The hemispherical shape is preferred to flat "windows" of material since radiation incident at an oblique angle on a flat material surface with index of refraction differing from air is subject to Fresnel losses [16], dependent on the angle of incidence of the radiation on the window. Hemispherical domes ensure the incidence angle of the beam radiation is always normal to the dome surface.

For pyranometers with a thermopile detector, the hot junctions are in contact with the flat, black, absorbing surface, and the "cold" reference junctions are shaded and perhaps in thermal contact with the case of the instrument, depending on the radiometer design. Alternatively, the flat receiving surface may consist of alternate black absorber and white reflecting segments in contact with the hot and cold thermopile junctions, respectively. These black and white detectors often have varying response sensitivity with respect to azimuth, depending on the pattern of white and black segments on the detector and the geometry with respect to the solar beam.

The sky dome has an effective temperature that is usually much lower than the ambient temperature at the surface [17]. This temperature difference results in an infrared (IR) energy exchange between the detector surface and the sky, usually with the detector emitting energy to the sky. This produces a negative contribution to the total signal of the pyranometer. This IR offset contribution varies with the variations in the sky dome and ambient temperatures, especially as modified by cloud cover and water vapor in the atmosphere. For the black-and-white detector design, both panels experience the same offset, so the relative error due to this effect is minimal. For an all-black detector, the IR exchange is entirely through the hot junctions, so there is a relatively larger error contribution for IR offset for these detectors [18].

The IR offset signals are most apparent as negative values of irradiance in night-time hours. The magnitude of these signals is about -1 Wm^{-2} to -3 Wm^{-2} for black-and-white designs and -5 Wm^{-2} to -20 Wm^{-2} for the all-black detectors. These errors contribute to the overall uncertainty in the pyranometer calibration and measurements.

2.4.2　Photoelectric Pyranometers

Photoelectric, or photodiode, pyranometer detectors suffer from Fresnel losses at the (generally) flat surface of the detectors themselves, as well as the surface of any flat window in the optical path to the detector. Therefore, these detectors are generally installed behind "diffusers," or materials that scatter or randomize the propagation direction of the photons while having good transmittance properties. Thus, the detectors "see" a relatively isotropic or homogeneous radiation field, and Fresnel loss effects are minimized. The geometrical or cosine response of the sensor is generally improved, as long as the spectral transmittance of the diffuser is not greater at some wavelengths versus others. This last effect would contribute an additional source of (spectral) uncertainty to the overall accuracy of these detectors.

As mentioned in Section 2.2.2, these detectors respond only to limited parts of the solar spectrum and cannot respond to or generate signals proportional to variations in those regions where the detector is "blind." This includes variations in the spectrum of the ground-reflected radiation, which usually has relatively large energy content in the regions where the detectors are blind (namely, the IR). This issue and the other limitations discussed in Section 2.2.2 result in the photoelectric pyranometers generally having larger uncertainty than thermal radiometers [19,20].

Thermal and photoelectric detectors each respond to different spectral regions. Typical spectral response curves for thermopile and silicon photodiode radiometer sensors are shown in Figure 2.3. Spectral distributions of solar energy are discussed in Section 2.5. These response limitations directly influence the uncertainty in the measured solar irradiance for each type of radiometer, as discussed briefly in Section 2.6.

Simple models for clear sky spectral distribution are discussed briefly in Chapter 7.

2.4.3　Diffuse Measurements

Diffuse sky radiation originates from the hemispherical sky dome and is measured directly using pyranometers mounted under devices that block the direct beam radiation. Such devices include disks or balls that track the sun, just as a pyrheliometer does. The tracking devices are designed to block the same FOV as the acceptance angle of a pyrheliometer. An alternative is to use a solid band of material positioned to obscure the entire path of the solar disk over the course of the day. Since the solar path changes location each day, the fixed shadow band must be adjusted to match these paths. The same is true of the tracking disk or ball system unless an automatic, computer-controlled positioning system is used.

Since the shadow band obscures the portion of the sky not occupied by the solar disk, a correction for the missing sky radiation is required. Often, the required corrections are a function of the sky condition (clear, partly cloudy, or overcast). The most popular shadowband correction is that of Drummond [21]. The application of

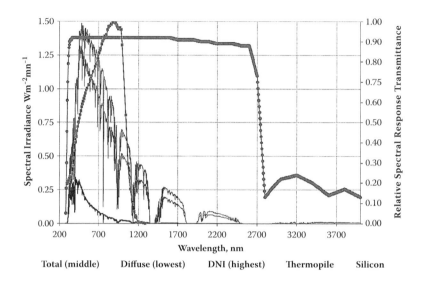

FIGURE 2.3 Spectral response of thermopile (circles) and silicon (squares) photodiode detectors.

the obscuration correction, with its inaccuracies, contributes to greater uncertainty in diffuse data using this technique versus the tracking shading disk technique.

Since the spectral distribution of diffuse radiation within the spectral response region of photoelectric detectors is extremely different from the global radiation, these detectors are much less accurate than thermal pyranometers for measuring diffuse radiation.

2.4.4 ROTATING SHADOWBAND RADIOMETERS

Given the basic component summation equation for solar radiation (Section 1.4), it is possible to measure all three components with a single radiometer or combination of any two radiometers. Thus, a pyrheliometer and pyranometer measuring global hemispherical or diffuse sky irradiance can be used to compute diffuse or global hemispherical radiation, respectively. Without a pyrheliometer, global and diffuse measurements can be used to compute the direct beam irradiance (Figure 2.4).

Given a single pyranometer, by alternately making measurements in a shaded and an unshaded condition, global and diffuse radiations are measured alternately, and the direct beam can be computed from these measurements [22]. These approaches require computation of the solar geometry. Since the geometry is constantly changing and the radiometers have finite response times, careful consideration to the timing of the measurements has to be considered.

Since photoelectric detectors have extremely fast time responses, several commercial varieties of rotating shadowband radiometers (RSRs) have been designed to measure all three solar components with one system (Figure 2.5). Extensive research has been conducted to develop corrections for spectral (especially for the shaded

FIGURE 2.4 (See color insert.) Left to right: Shaded, ventilated thermal pyranometer for diffuse sky irradiance; unshaded thermal pyranometer for total hemispherical irradiance; same model shaded thermal pyranometer and thermal pyrheliometer (below pyranometers) mounted on a solar tracker.

FIGURE 2.5 (See color insert.) Rotating shadow-band radiometer. Rotating band is below silicon photodiode photoelectric pyranometer (just above horizon line).

diffuse measurements), geometrical, and temperature response of the pyranometers. However, no amount of correction schemes can reduce the basic uncertainty in the pyranometer device for routine monitoring applications under all conditions [23,24]. Only if very detailed and expensive spectral and temperature measurements can be made in conjunction with the broadband radiometer data can some characterization of or improvement in the accuracy of the RSR type of data be made.

2.5   SPECTRAL DISTRIBUTIONS

Electromagnetic emissions from the sun extend across the electromagnetic spectrum from the highly energetic x-ray region through the ultraviolet, visible, and IR portion of the spectrum to the far IR and radio region. These emissions interact with Earth's own electromagnetic and atmospheric envelope, resulting in large variations in the magnitude of solar radiation available for conversion into other forms of useful energy. Most solar energy conversion systems utilize only part of the solar spectrum. For example, in lighting, the energy in the human visual response range (wavelengths of 380 to 850 nm) are important. Crystal silicon photovoltaic cells convert radiation from 350 to 1100 nm, and plants use photosynthetically active radiation (PAR) between 400 and 700 nm [25]. However, the efficiency and performance of these systems are usually based on the total solar broadband radiation measured between 250 and 2500 nm. Specialized instruments and measurement techniques have been developed to quantify the available broadband solar resource and for spectral regions utilized by each technology, shown at the top of Figure 2.6.

The spectra in Figure 2.6, except for the extraterrestrial example, are only single examples of each of the component spectra. The spectral distribution of each component changes as air mass changes and as the constituents of the atmosphere change. Figures 2.7 and 2.8 show how the direct and global spectral distributions change as a

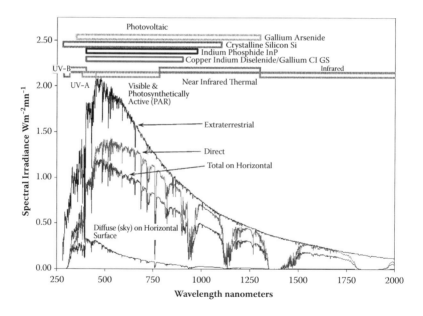

FIGURE 2.6 (See color insert.)   Representative distributions of power as a function of wavelength for solar energy at the top of the atmosphere (extraterrestrial), solar beam radiation, diffuse sky radiation, and total radiation on a horizontal surface for clear sky conditions. Bars show regions of energy utilization by conversion systems. Note especially the difference between the global and diffuse horizontal spectra.

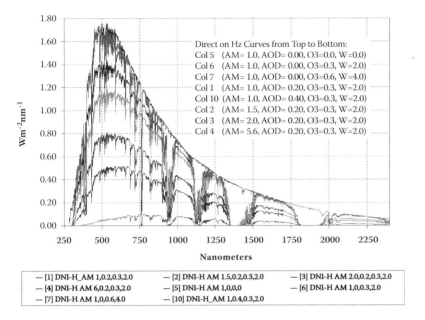

FIGURE 2.7 Direct normal spectral distributions on a horizontal surface as a function of air mass, aerosol optical depth, ozone, and water vapor.

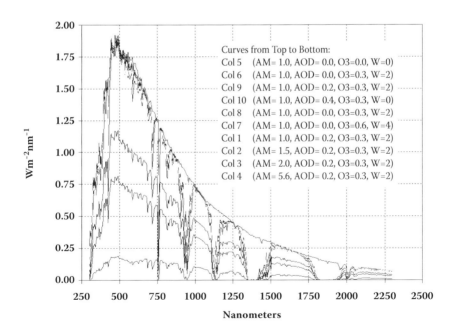

FIGURE 2.8 Global hemispherical spectral distributions on horizontal surface as a function of air mass, aerosol optical depth, ozone, and water vapor.

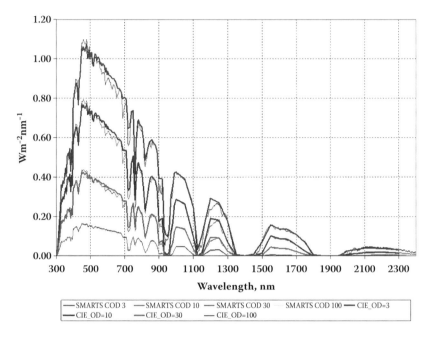

FIGURE 2.9 Global hemispherical spectral distributions for various overcast cloud optical depths.

function of air mass (AM), aerosol optical depth (AOD), ozone (O_3), and total water vapor (w) content. Discussion of spectral models can be found in the work of Bird and Riordan [26] and Gueymard [27].

Under cloudy conditions, the clouds introduce their own attenuation, often referred to as cloud optical depth. The type and physical configuration of clouds can be highly variable. Figure 2.9 shows the impact of increasing cloud optical depth for overcast skies, based on cloud attenuation models developed by Justus and Paris [28].

2.6   UNCERTAINTY

A detailed discussion of the propagation of uncertainty in various measurement devices is not the object of this book. But, a description of the concepts involved and basic results is in order. Uncertainty analysis is the derivation of the possible error in a measurement instrument or data point derived from an instrument. Uncertainty consists of precision (repeatability, scatter of data) and accuracy (how different in magnitude and direction) with respect to the exact, correct, or "true" value of the measurement. The accuracy of models cannot be greater than the uncertainty in the data used to validate the models. Models developed using correlations between input parameters and irradiance measurements inherently carry the uncertainty in the measurements into the models.

2.6.1 THE GUIDE TO MEASUREMENT UNCERTAINTY

There is a standard procedure for evaluating the uncertainty in measurements, developed by the international consensus documentary standards community. This is known informally as the "Guide to Uncertainty in Measurement" or GUM [29]. Several publications are available describing application of these principles to solar radiometer calibrations and measurements. Basically, all the contributing elements to uncertainty are identified regarding their source, type (type A: statistical or "random"; type B: nonstatistical or "systematic"), statistical distribution, and magnitude. These are then combined using specific rules to produce an overall "expanded uncertainty" with a reported confidence interval.

A measurement equation describing the process to be analyzed (such as an equation describing the derivation of a calibration factor from measured parameters) is required. Estimates or derived standard uncertainties are assigned to each variable in the measurement equation. Standard uncertainty Us is representative of the statistical standard deviation of Gaussian or normally distributed data for the response, especially from empirical data that can be analyzed with statistical procedures. The standard uncertainty is dependent on the distribution of the response data being analyzed. The standard uncertainty may need to be derived or estimated from specifications, independent test data, or engineering experience. If functions or equations are fit or used in the measurement or calibration process, the standard errors of estimates or mean residuals from fitted functions serve as additional sources of standard uncertainty in the process.

The sensitivity of the result to variations in each of the measured parameters is derived by computing the sensitivity coefficient c_i for the ith parameter, the partial derivative of the response with respect to each measurement equation variable x_i: $c_i = \partial y/\partial x_i$. Inserting values of the measured parameters into the equations for the sensitivity coefficients produces estimates of error contributions due to each measurement variable for each measurement point when the c_i values are multiplied by the estimated standard uncertainty in each measurement parameter.

Combined standard uncertainty Uc is computed from the root sum square of the products c_i Ux_i (and possibly other sources of standard uncertainty, such as from regression equations). Last, depending on the known or assumed statistical distribution type for the data, a "coverage factor" k is selected. This factor, multiplied by the combined standard uncertainty Uc, produces an expanded uncertainty. The combination of the choice of data distribution shape and confidence level (95%, 99%, etc.) will dictate the value of k. For Gaussian distributed data and confidence level of 95%, k = 2 is appropriate. See the GUM [29] or Reda et al. [30] and Reda [31] for detailed examples.

2.6.2 SOURCES OF RADIOMETRIC UNCERTAINTY

Sources of radiometric uncertainty contributing to the overall uncertainty in a result can be categorized as originating from three sources:

- Characterization (calibration) of measurement instrumentation
- Data acquisition and recording equipment
- Data reduction and processing (mathematical manipulation or modeling)

Calibration and characterization sources of uncertainty relate to the influences of various parameters on the radiometers as briefly mentioned. These include spectral response, temperature response, IR offsets, and so on, as well as the accuracy of the reference standard for calibrating the radiometers.

Data acquisition-related sources include the uncertainty in measurement of sensor signals and environmental influences and limitations of electronic measurement systems. These are usually prescribed in instrumentation specifications and require careful interpretation.

Uncertainty due to data reduction processing results from the distribution and scatter of data about correlation lines, sample sizes for averaging, and other statistical and mathematical properties derived in computing a result from an assemblage of data (which are themselves uncertain).

Brief descriptions of the solar radiometer reference standard and one of the most important instrument characteristics established during calibration (geometrical response function for pyranometers) are given in the next two sections.

2.6.3 THE WORLD RADIOMETRIC REFERENCE

In the early twentieth century, solar radiometry calibrations were based on calorimetry, such as with the Smithsonian Institution silver disk pyrheliometers [32] and water flow pyrheliometers. Temperature differences between material alternately shaded and then unshaded from the solar beam established the magnitude of the direct beam energy or power content and solar measurement scales. Coulson [2] gave a detailed history of the development of solar radiation scales.

Beginning in 1970, Kendall [33], and later Willson [34], at the Jet Propulsion Laboratory developed electrical compensation radiometers utilizing conical blackened-silver cavities. The cavities were instrumented with thermojunctions that could be calibrated using electrical heating in place of heating from solar radiation. Instruments of similar design by Crommelynck [35], Brusa and Fröhlich [36], Hickey and Karoli [37], and Hickey et al. [38] were developed and have become the standard for the World Radiometric Reference (WRR) established by the World Meteorological Organization/World Radiation Center (WMO/WRC) and now in use [11]. Figure 2.10 is a photograph of a typical reference absolute cavity pyrheliometer.

Traceability of measurements requires an unbroken chain of comparisons with defined uncertainties to stated references. The traceability of the Working Standard Group (WSG) cavity radiometers to Systeme Internationale (SI) units is with respect to electrical (volt, ohm, and ampere) and physical-dimensional (length, area) standards maintained by national standardizing laboratories. Detailed characterization of the aperture area, absorption of the cavity, and electrical components of the measurement system substantiates the "absolute" nature of the WSG measurements.

The WRR is the measurement standard for the International System or SI unit of solar irradiance [39]. The WRR was introduced to ensure homogeneity of solar

FIGURE 2.10 (See color insert.)   Typical absolute cavity pyrheliometer for reference DNI measurements.

radiation measurements and has been in use since 1980 [11]. The stated uncertainty in the WRR determination is 0.3% for DNI greater than 700 Wm^{-2}. This value should be remembered as the absolute smallest possible measurement uncertainty for the direct beam or any solar radiation component. Comparisons have shown that the PMO6 model number WMO cavity radiometer (corrected to WRR) reproduced the SI laboratory irradiance scale to within 0.05% [40].

WMO sponsors International Pyrheliometer Comparisons (IPC) every 5 years. WMO representatives of any country and radiometer manufacturers are invited to bring their reference solar radiometers to Davos to compare with the WSG and derive a WRR calibration factor [11]. The test radiometers are compared with the WSG using a rigorous data collection and processing protocol described by Reda [41].

2.6.4 PYRANOMETER GEOMETRICAL RESPONSE FUNCTIONS

The density of the photons per unit area on the flat surface of the pyranometer detector depends on the angle between the normal to the surface and the incident photon. At low incidence angles, the areal density is greater; at large incidence angles, the density is lower. The Lambert cosine law reflects the dependence of the radiation, or flux, density:

$$I = Io \cos(\theta) \qquad (2.3)$$

where Io is the incident radiation intensity along the direction of propagation, θ is the incidence angle (between the surface normal and the direction of propagation of

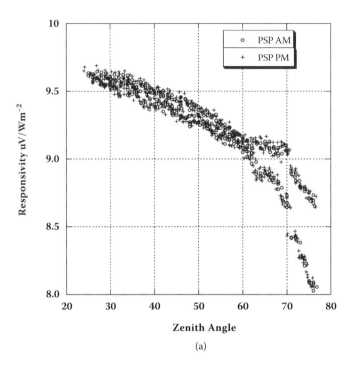

(a)

FIGURE 2.11   Pyranometer (a-above) and pyrheliometer (b-opposite) response versus zenith angle. Pyranometer response range is ±4% about the 45° Z value. Pyrheliometer response range is ±1.0% about the 45° Z value. Every instrument has a different "signature." (Source: NREL BORCAL reports available online. http://www.nrel.gov/aim/borcal.html.)

the photon), and I is the radiation flux at the detector absorbed by the detector that produces the signal. Thus, the DNI contribution to a pyranometer signal is B cos(θ) for a beam intensity of B. The remaining contribution to the pyranometer signal is the sky radiation on a horizontal surface, or diffuse horizontal irradiance (DHI). If the pyranometer is tilted, an additional contribution from ground-reflected radiation R will be intercepted by the detector.

Since no material is a perfect radiation absorber, each material or detector structure has its own individual deviations from the Lambert cosine law. If a detector were perfect, plotting the response of the detector as a function of the angle of incidence of the beam radiation would result in a flat, straight line. Real solar radiation detectors all display some deviation from this ideal response. The deviations are a combination of surface mechanical, optical, and electromagnetic properties and the IR offset for thermal detectors (discussed in Section 2.3.1 and [42]). Examples are given in Figure 2.11(a) for a pyranometer and 2.11(b) for a pyrheliometer. Note in particular the wide variation in cosine response for the pyranometer with respect to zenith or incidence angles. The difference between morning (a.m.) and afternoon (p.m.) pyranometer response may be the result of the sensor not being level or parallel to local horizontal. Pyrheliometer variations may be due to environmental conditions (wind, temperature) or changes in spectral content of the DNI from morning to noon to afternoon.

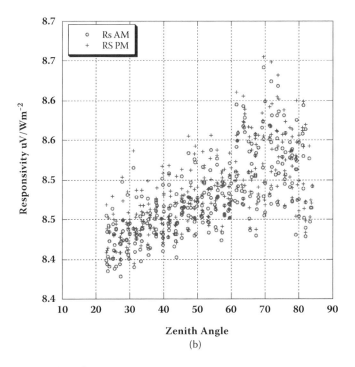

FIGURE 2.11 *(Continued)*

Thirty-five years of my experience in the calibration of solar radiometers has shown that each individual radiometer has its own individual, characteristic signature for this important response function. The data with the highest accuracy for solar radiometric data can only be achieved by correcting for these deviations. Using average response or response at specific incidence angles can produce relatively accurate aggregated data (averaged over a day, month, year, or longer periods). This is because many response "error" functions are somewhat symmetrical about noon or the minimum zenith angle and can "cancel out" in the averaging process. However, individual, instantaneous, or short-term pyranometer (GHI) data can often approach an uncertainty of 10% or greater in morning or afternoon data. These response functions can also become important for correcting radiometric data measured for determining solar energy conversion system performance, or efficiency, especially for solar collectors or systems configured as tilted systems. In these cases, the measurement configuration and geometry can be significantly different from the calibration conditions (typically horizontal).

2.6.5 Summary of Uncertainty Sources and Magnitudes in Solar Measurements

The introduction to solar radiation measurements described the instrumentation and some sources of uncertainty in solar radiation measurements. Tables 2.1 and 2.2 tabulate the main sources of uncertainty for pyrheliometer and pyranometer calibrations. Measured data uncertainty starts with these uncertainties but includes

TABLE 2.1

Uncertainties for Individual Pyrheliometer Calibrations (%)

Type A (Statistical)	Standard Uncertainty, Ux_i	U95 Uncertainty, $k = 2 \times Ux_i$	Type B (Nonstatistical)	Standard Uncertainty, Ux_i	U95 Uncertainty, $k = 2 \times Ux_i$
WRR transfer	0.100	0.200	WRR uncertainty	0.150	0.300
Temperature response	0.250	0.500	Temperature ±10°C	0.125	0.250
Data logger precision	0.003	0.005	Data logger bias	0.045	0.090
Linearity	0.100	0.200	Field of view	0.025	0.050
Cavity reference DNI	0.013	0.025	Cavity reference DNI	0.013	0.025
Tracking	0.125	0.250	Tracking bias	0.013	0.025
Spectral	0.250	0.500	Spectral error	0.250	0.500
Thermal EMF (electromotive force)	0.003	0.005	Thermal EMF	0.003	0.005
Electromagnetic interference	0.003	0.005	Electromagnetic interference	0.003	0.005
Total A	0.401	0.802	Total B	0.322	0.644
Combined uncertainty	1.03%				
Expanded uncertainty	2.06%				

TABLE 2.2
Uncertainties in Individual Thermopile Pyranometer Calibration (%)

Type A (Statistical)	Standard Uncertainty, Ux_i	U95 Uncertainty, $k = 2 \times Ux_i$	Type B (Nonstatistical)	Standard Uncertainty, Ux_i	U95 Uncertainty, $k = 2 \times Ux_i$
WRR transfer	0.100	0.200	WRR uncertainty	0.150	0.300
Cos(Z); 2° interval	0.250	0.500	Cos(Z) bias	0.005	0.010
				(1.5%)	(3.0%)
Diffuse variation	0.003	0.005	Diffuse offset	0.063	0.125
Temperature	0.100	0.200	Logger bias	0.045	0.090
Logger precision	0.013	0.025	Cavity bias	0.013	0.025
Cavity precision	0.125	0.250	Cavity track bias	0.013	0.025
Spectral variation	0.250	0.500	Spectral bias	0.250	0.500
Thermal EMF	0.003	0.005	Thermal EMF bias	0.003	0.005
Test IR Offset	0.125	0.250	Temperature bias	0.125	0.250
Electromagnetic interference	0.003	0.005	Electromagnetic interference	0.003	0.005
Regression Standard Error of Estimate (SEE)	0.250	0.500			
Total A	0.489	0.978	Total B	0.327	0.654
Combined uncertainty			1.2% (3.2%)		
Expanded uncertainty			2.4% (6.4%)		

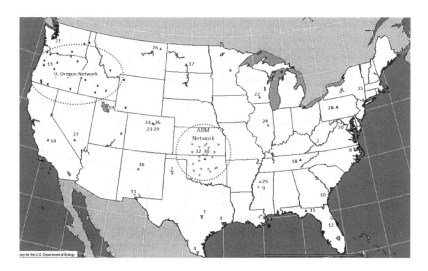

FIGURE 2.13 Principal research quality solar irradiance measurement stations and networks in the United States as of 2012. See Table 2.3 for station listing. (Adapted from Renné, D., T. Stoffel, M. Anderberg, P. Gray-Hann, and J. Augustyn. 2000. Current status of solar measurement programs in the U.S. Campbell-Howe, R., ed. *Proceedings of the Solar 2000 Conference 16–21 June 2000, Madison, Wisconsin.* NREL/CP-560-28115. American Solar Energy Society, Boulder, CO, and map of U.S. solar measurement station locations at http://www.nrel.gov/gis/tools.html. Accessed 14 August 2012.)

The World Radiation Data Center (WRDC) in St. Petersburg, Russia, is the WMO repository for measured solar radiation data ranging from sunshine duration to global hemispherical, direct normal, and diffuse horizontal radiation. Each WRDC station has a different complement of sensors and period of record. The data are reported in units of percentage sunshine or joules per square centimeter. Some stations report hourly averages, but the most commonly reported data are daily total radiation. Figure 2.14 is a map of the WRC stations for which at least one month of data is archived.

From the WRDC Web site, the following description of how daily and monthly data averages and totals are computed is summarized and gives an indication of filtering required to compute meaningful statistics:

1. The daily total of radiation balance is computed for all 24-h intervals available. If at least one hourly total value is missing, the daily total is not calculated. For a missing data symbol, "-" is entered in the table instead.
2. In computing daily totals, sunrise and sunset hour intervals are specially treated to avoid undue reduction in daily totals due to the gaps in the data for those hours:
   If observational data are missing for the mentioned hour intervals on a number of days, the solar elevation angle e is computed for those hours.
   If e > 5°, the missing hourly total is flagged as missing, and the daily total is not computed. If e ≤ 5°, the daily total is calculated (estimated) based on the partial hour.

TABLE 2.3
List of Station Locations in Figure 2.13

Station	Location	Network
1	University of Texas Austin, Austin, TX	NREL CONFRRM[a]
2	West Texas A&M University, Canyon, TX	NREL CONFRRM
3	Johnson Space Flight Center, Houston, TX	NREL CONFRRM
4	University of Texas Pan American, Edinburgh, TX	NREL CONFRRM
5	University of Texas El Paso, El Paso, TX	NREL CONFRRM
6	Bethune-Cookman, Daytona Beach, FL	NREL HBCU[b]
7	Bluefield State College, Bluefield, WV	NREL HBCU
8	Elizabeth City State College, Elizabeth City, NC	NREL HBCU
9	Mississippi Valley State, Itta Bena, MS	NREL HBCU
10	Savannah St. College, Savannah, GA	NREL HBCU
11	Las Cruces, NM	NREL CONFRRM
12	Cocoa, FL	NREL CONFRRM
13	Eugene, OR (and University of Oregon Network)	NREL CONFRRM
14	Hanford, CA	NOAA ISIS[c]
15	Oak Ridge, TN	NOAA ISIS
16	Albuquerque, NM	NOAA ISIS
17	Bismarck, ND	NOAA ISIS
19	Salt Lake City, UT	NOAA ISIS
20	Sterling, VA	NOAA ISIS
21	Seattle, WA	NOAA ISIS
22	Madison, WI	NOAA ISIS
23	Boulder, CO	NOAA SURFRAD[d]
24	Bondville, IL	NOAA SURFRAD
25	Goodwin Creek, MS	NOAA SURFRAD
26	Fork Peck, MT	NOAA SURFRAD
27	Desert Rock, NV	NOAA SURFRAD
28	Pennsylvania State University, PA	NOAA SURFRAD
29	Erie, CO	NOAA Boulder (BOA)[e]
30	Southern Great Plains (SGP), Kansas/Oklahoma	DOE ARM program[f]
32	Lamont, OK	NOAA SURFRAD
33	Golden, CO[g]	NREL
35	Albany, NY	International Daylight Monitoring Program[h]
36	Aurora, CO[i]	Solar Technology Acceleration Center

[a] http://rredc.nrel.gov/solar/new_data/confrrm/#map.
[b] http://rredc.nrel.gov/solar/old_data/hbcu
[c] http://www.srrb.noaa.gov/isis/isissites.html.
[d] http://www.srrb.noaa.gov/surfrad/sitepage.html.
[e] http://www.esrl.noaa.gov/psd/technology/bao.
[f] http://www.arm.gov/sites/sgp.
[g] http://www.nrel.gov/midc.
[h] http://idmp.entpe.fr.
[i] http://www.nrel.gov/midc/solartac.

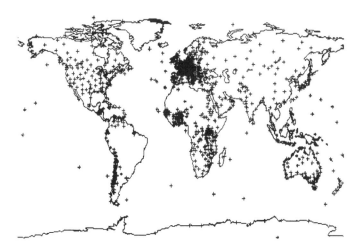

FIGURE 2.14 Map of WMO stations reporting data to the World Radiation Center, St. Petersburg, Russia. (From http://wrdc-mgo.nrel.gov/html/mapap.html. See also http://wrdc. mgo.rssi.ru.)

3. If data for each day of the month are available,
   a. The monthly total is calculated by adding together daily/hourly totals for each day of the month;
   b. The monthly mean of daily/hourly totals is calculated by dividing the monthly total by the number of days in the calendar month.
4. If there are days with missing data, the monthly mean of daily/hourly totals from the existing data is calculated, and the monthly total is computed from the following:
   a. The monthly mean of daily/hourly totals is calculated by dividing the sum of daily/hourly totals for the days with available data (including estimated and questionable values) by the number of days with observations;
   b. The monthly total is calculated by multiplying the monthly mean of daily/hourly totals by the number of days in the calendar month.
5. If there are 10 or more days with missing daily/hourly totals, the monthly total and the monthly mean of daily/hourly totals are not calculated, and the missing data flag "-" is entered in the respective columns.
6. If there are from 5 to 9 days with missing daily/hourly totals, the monthly total and the monthly mean of daily/hourly totals are enclosed in brackets.
7. If there are five or more daily/hourly totals enclosed in brackets (estimated or questionable), the monthly totals and the monthly mean of daily/hourly totals are enclosed in brackets.
8. If the number of days with missing daily/hourly totals and daily/hourly totals in brackets added together is from 5 to 9, the monthly total and the monthly mean of daily/hourly totals are enclosed in brackets, and if there are 10 or more such days, the monthly total and the monthly mean of daily/ hourly totals are not calculated; "-" is entered in respective columns. All radiation parameters and sunshine duration are subject to this rule.

9. The monthly means of daily totals of appropriate radiation parameters along with monthly means of hourly totals are computed based on the available data and the rules as given here.

Data can be accessed and downloaded from http://wrdc.mgo.rssi.ru (last accessed 26 Oct 2012). The data are carefully assessed for quality by visual inspection and automated filtering software. For more details on the WRDC data products, see Tsvetkov, Wilcox, Renné, and M. Pulscak [47] and Driesse and Thevenard [48].

2.8 SUMMARY

There are many measurement principles and approaches to measuring solar radiation. These measurements are some of the least-accurate physical measurements. While other physical quantities such as mass, electric current, and the like can be measured with accuracies of parts per million or even better, our best present instrumentation for solar measurements is no more accurate than about 3 parts per thousand, or 0.3%. Our best routine measurements are no more accurate than a few parts per hundred. This is due to the difficulty in physically collecting and converting photons into some electrical or thermal signal that can be easily measured. Quality measurements of solar radiation are rare. This is mainly because of the expense of the instrumentation but also largely because of the labor involved to calibrate and maintain the instruments and collect, assess, process, and archive the data. Even though the quality of future civilization may depend on the balance of incoming and outgoing solar radiation, high-quality measurements are poorly distributed around the world. Engineering of solar energy systems based on these measurements is fraught with uncertainty that grows as fewer and lower-quality measurements are used. Last, when computed or modeled values of solar radiation are used to design solar energy conversion systems, it is imperative to remember that no model is any more accurate than the measured data that can be used to develop or validate the model.

With this introductory material in mind, the next chapter begins the discussion of modeling solar radiation under clear, or rather better to say, cloudless skies.

REFERENCES

1. Vignola, F., J. Michalsky, and T. Stoffel. (2012). *Solar and Infrared Radiation Measurements*. CRC Press, Taylor and Francis Group, Boca Raton, FL.
2. Coulson, K. (1975). *Solar and Terrestrial Radiation*. Academic Press, New York.
3. Abbot, C.G., and L.B. Aldrich. (1932). *An Improved Water-Flow Pyrheliometer and the Standard Scale of Solar Radiation*. Smithsonian Miscellaneous Collections 87, No. 15. Smithsonian Institution, Washington, DC.
4. Seebeck, T.J. (1821). *Ueber den magnetismus der galvenische kette*. Abhandlungen der Koniglich Akadamie der Wissenshaften, Berlin, pp. 289–325.
5. Benedict, R. (1981). *Fundamentals of Temperature, Pressure and Flow Measurements*, 3rd ed. Wiley, Hoboken, NJ.

6. Burns, G.W., M. Scroger, G.F. Strouse, M.C. Croarkin, and W.F. Guthrie (1993). *Temperature-Electromotive Force Reference Functions and Tables for the Letter-Designated Thermocouple Types Based on the ITS-90*. NIST Monograph 175. U.S. Department of Commerce Technology Administration, National Institute of Standards and Technology, Gaithersburg, MD. See http://srdata.nist.gov/its90/main/. Accessed 21 July 2012.

7. Tew, W.L., and G.F. Strouse. (2001). *Standard Reference Material 1750: Standard Platinum Resistance Thermometers, 13.8033 K to 429.7485 K*. NIST Special Publication 260-139. U.S. Department of Commerce Technology Administration, National Institute of Standards and Technology, Gaithersburg, MD.

8. Steinhart, I.S., and S.R. Hart. (1968). Calibration curves for thermistors. *Deep Sea Research*, Vol. 15, p. 497.

9. Einstein, A. (1905). On the thermodynamic theory of the difference in potentials between metals and fully dissociated solutions of their salts and on an electrical method for investigating molecular forces. *Annalen der Physik,* Ser. 4, 8, pp. 798–814.

10. Pankove, J. (ed.). (1971). *Optical Processes in Semiconductors*. Prentice Hall, Englewood Cliffs, NJ.

11. World Meteorological Organization. (2008). *WMO Guide to Meteorological Instruments and Methods of Observation,* WMO No. 8, 7th ed. Geneva, Switzerland. http://www.wmo.int/pages/prog/gcos/documents/gruanmanuals/CIMO/CIMO_Guide-7th_Edition-2008.pdf, Accessed 21 July 2012.

12. van de Hulst, H.C. (1957). *Light Scattering by Small Particles*. Wiley, New York.

13. Neumann, A., A. Witzke, S.A. Jones, and G. Schmitt. (2002). Representative terrestrial solar brightness profiles. *Journal of Solar Energy Engineering, Transactions of the ASME*, Vol. 124, p. 198.

14. Buie, D., and A.G. Monger. (2004). The effect of circumsolar radiation on a solar concentrating system. *Solar Energy*, Vol. 76, pp. 181–185.

15. Mie, G. (1908). Beiträge zur Optik trüber Medien, speziell kolloidaler Metallösungen. *Annalen der Physik,* Vol. 330, pp. 377–445.

16. Hecht, E. (2001). *Optics,* 4th ed. Addison-Wesley, Boston.

17. Jursa, A.S. (1985). *Handbook of Geophysics and the Space Environment*. Air Force Geophysics Laboratory, Air Force Systems Command, United States Air Force, Hanscom Air Force Base, MA. PDF available online at http://www.dtic.mil/cgi-bin/GetTRDoc?Location=U2&doc=GetTRDoc.pdf&AD=ADA167000 (accessed 26 Oct 2012).

18. Dutton, E.G., J.J. Michalsky, T. Stoffel, B.W. Forgan, J. Hickey, D.W. Nelson, T.L. Alberta, and I. Reda. (2001). Measurement of broadband diffuse solar irradiance using current commercial instrumentation with a correction for thermal offset errors. *Journal of Atmospheric and Ocean Technology*, Vol. 18, pp. 219–314.

19. King, D.L., and D.R. Myers. (1997). Silicon-photodiode pyranometers: Operational characteristics, historical experiences, and new calibration procedures. *Proceedings of the 26th IEEE Photovoltaic Specialists Conference*, September 29–October 3, 1997, Anaheim, California. Institute of Electrical and Electronics Engineers.

20. Myers, D. (2011). Quantitative analysis of spectral impacts on silicon photodiode radiometers. *Proceedings of SOLAR 2011*, May 17–21, 2011, Raleigh, North Carolina. American Solar Energy Society, Boulder, CO.

21. Drummond, A.J. (1956). On the measurement of sky radiation. *Theoretical and Applied Meteorology*, Vol. 7, No. 3–4, pp. 413–436.

22. Michalsky, J.J., J.L. Berndt, and G.J. Schuster. (1986). A microprocessor-based rotating shadowband radiometer. *Solar Energy*, Vol. 36, No. 5, pp. 465–470.

23. Michalsky, J.J., L. Harrison, and B.A. LeBaron. (1987). Empirical radiometric correction of a silicon photodiode rotating shadowband pyranometer. *Solar Energy*, Vol. 39, No. 2, pp. 87–96.

24. Michalsky, J.J., L. Harrison, and B.A. LeBaron. (1991). Spectral and temperature correction of silicon photovoltaic solar radiation detectors. *Solar Energy*, Vol. 47, No. 4, pp. 299–305.

25. Biggs, W.W., A. R. Edison, J. D. Eastin, K.W. Brown, J.W. Maranvile, and M.D. Clegg. (1971). Photosynthesis light sensor and meter. *Ecology*, Vol. 52, pp. 125–131.

26. Bird, R.E., and C. Riordan. (1986). Simple solar spectral model for direct and diffuse irradiance on horizontal and tilted planes at the earth's surface for cloudless atmospheres. *Journal of Climate and Applied Meteorology*, Vol. 25, pp. 87–97.

27. Gueymard, C., (2001). Parameterized transmittance model for direct beam and circumsolar spectral irradiance. *Solar Energy*, Vol. 71, No. 5, pp. 325–346.

28. Justus, C.G., and M.V. Paris (1988). A cloudy sky radiative transfer model suitable for calibration of satellite sensors. *Remote Sensing of the Environment*, Vol. 24, pp. 269–285.

29. Working Group 1 of the Joint Committee for Guides in Metrology (JCGM/WG 1). (2008). *Evaluation of measurement data—Guide to the expression of uncertainty in measurement JCGM 100. GUM 1995 with minor corrections 2008.* http://www.bipm.org/utils/common/documents/jcgm/JCGM_100_2008_E.pdf. Accessed 21 July 2012.

30. Reda, I., and D. Myers, T. Stoffel. (2008). Uncertainty estimate for the outdoor calibration of solar pyranometers: A metrologist perspective. Measure. NREL Report No. JA-581-41370. *NCSLI Journal of Measurement Science*, Vol. 3, No. 4, December 2008, pp. 58–66.

31. Reda, I. (2011). *Method to Calculate Uncertainties in Measuring Shortwave Solar Irradiance Using Thermopile and Semiconductor Solar Radiometers.* NREL Technical Report NREL/TP-3B10-52194. http://www.nrel.gov/docs/fy11osti/52194.pdf. Accessed 20 July 2012.

32. Abbot, C.G. (1911). *The Silver Disk Pyrheliometer.* Smithsonian Miscellaneous Collections—Vol. 56, No. 19. Smithsonian Institution, Washington, DC.

33. Kendall, J.M., and C.M. Berdahl. (1970). Two blackbody radiometers of accuracy. *Applied Optics*, Vol. 9, No. 5, pp. 1082–1091.

34. Willson, R.C. (1973). Active cavity radiometer. *Applied Optics*, Vol. 12, No. 4, pp. 810–817.

35. Crommelynck, D. (1977). Calibration of radiation instruments for the measurement of the radiant flux of an arbitrary source. *Applied Optics*, Vol. 16, No. 2, p. 302.

36. Brusa, R.W., and C. Fröhlich. (1986). Absolute radiometers (PMO6) and their experimental characterization. *Applied Optics*, Vol. 25, pp. 4173–4180.

37. Hickey, J.R., and A.R. Karoli. (1974). Radiometric calibrations for the Earth Radiation Budget experiment. *Applied Optics*, Vol. 13, pp. 523–533.

38. Hickey, J.R., L.L. Stowe, H. Jacobowitz, P. Pellegrino, R.H. Maschhoff, F. House, and T.H. Vonder Harr. (1980). Initial solar irradiance determinations from Nimbus 7 cavity radiometer measurements. *Science*, Vol. 208, no. 4441, pp. 281–283.

39. Rüedi, I., and W. Finsterle. (2005). *The World Radiometric Reference and Its Quality System.* Technical Conference on Meteorological and Environmental Instruments and Methods of Observation—TECO 2005. http://www.wmo.int/pages/prog/www/IMOP/publications/IOM-82-TECO_2005/Papers/3(15)_Switzerland_Ruedi.pdf. Accessed 21 July 2012.

40. Romero J., N.P. Fox, and C. Frohlich. (1996). Improved comparison of the World Radiometric Reference and the SI radiometric scale. *Metrologia*, Vol. 32, No. 6, pp. 523–524.

41. Reda, I. (1996). *Calibration of a Solar Absolute Cavity Radiometer with Traceability to the World Radiometric Reference.* NREL Technical Report/TP-463-20619, National Renewable Energy Laboratory, Golden, CO. 79 pp. http://www.nrel.gov/rredc/pdfs/20619.pdf. Accessed 21 July 2012.

42. Myers, D.R., T.L. Stoffel, I. Reda, S.M. Wilcox, and A.M. Andreas. (2002). Recent progress in reducing the uncertainty in and improving pyranometer calibrations. *Journal of Solar Energy Engineering, Transactions of the ASME*, Vol. 124, pp. 44–50.

43. Stoffel, T., D. Renné, D. Myers, S. Wilcox, M. Sengupta, R. George, and C. Turchi. (2010). *Concentrating Solar Power Best Practices Handbook for the Collection and Use of Solar Resource Data*. NREL Technical Report NREL/TP-550-47465. http://www.nrel.gov/docs/fy10osti/47465.pdf. Accessed 9 July 2012.

44. Myers, D.R. (2006). The necessity and economics of solar radiation resource assessment. Campbell-Howe, R., ed. *Proceedings of the Solar 2006 Conference* 7–13 July 2006, Denver, CO. American Solar Energy Society, Boulder, CO.

45. Renné, D., T. Stoffel, M. Anderberg, P. Gray-Hann, and J. Augustyn. (2000). Current status of solar measurement programs in the U.S. Campbell-Howe, R., ed. *Proceedings of the Solar 2000 Conference 16–21 June 2000, Madison, Wisconsin*. NREL/CP-560-28115. American Solar Energy Society, Boulder, CO.

46. Hall, J., and J. Hall. (2011). Quality analysis of global horizontal irradiance data from 3500 U.S. ground-based weather stations. *Proceedings of the Solar 2011 Conference*, 17–20 May 2011, Rayleigh, NC. American Solar Energy Society, Boulder, CO. http://www.solardatawarehouse.com. Accessed 9 July 2012.

47. Tsvetkov, A., S. Wilcox, D. Renné, and M. Pulscak. (1995). International solar resource data at the World Radiation Data Center. In *Proceedings 1995 American Solar Energy Society Annual Conference,* Minneapolis MN, pp. 216–219. American Solar Energy Society, Boulder, CO.

48. Driesse, A., and D. Thevenard (2002), A test of Suehrcke's sunshine radiation relationship using a global data set. *Solar Energy*, Vol. 72, No. 2, pp. 167–175.

FIGURE 2.2 Thermal (top right) and solid-state photoelectric detector pyranometers including a silicon photovoltaic reference cell (bottom left).

FIGURE 2.4 Left to right: Shaded, ventilated thermal pyranometer for diffuse sky irradiance; unshaded thermal pyranometer for total hemispherical irradiance; same model shaded thermal pyranometer and thermal pyrheliometer (below pyranometers) mounted on a solar tracker.

FIGURE 2.5 Rotating shadow-band radiometer. Rotating band is below silicon photodiode photoelectric pyranometer (just above horizon line).

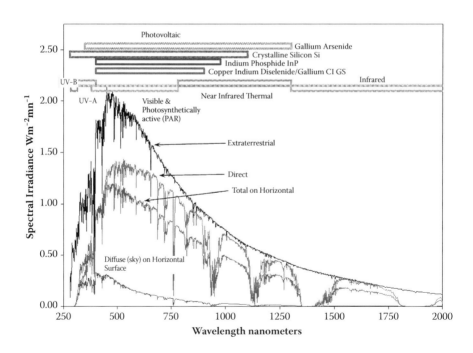

FIGURE 2.6 Representative distributions of power as a function of wavelength for solar energy at the top of the atmosphere (extraterrestrial), solar beam radiation, diffuse sky radiation, and total radiation on a horizontal surface for clear sky conditions. Bars show regions of energy utilization by conversion systems. Note especially the difference between the global and diffuse horizontal spectra.

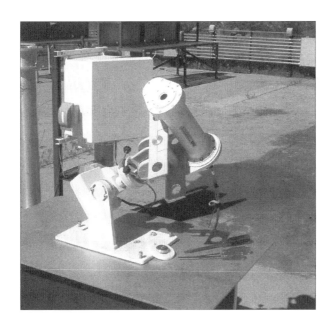

FIGURE 2.10 Typical absolute cavity pyrheliometer for reference DNI measurements.

FIGURE 6.1 Global hemispherical pyranometers (left, silicon photodiode; right, thermopile) measuring plane of array irradiance for photovoltaic performance testing.

FIGURE 2.2 Thermal (top right) and solid-state photoelectric detector pyranometers including a silicon photovoltaic reference cell (bottom left).

FIGURE 2.4 Left to right: Shaded, ventilated thermal pyranometer for diffuse sky irradiance; unshaded thermal pyranometer for total hemispherical irradiance; same model shaded thermal pyranometer and thermal pyrheliometer (below pyranometers) mounted on a solar tracker.

FIGURE 2.5 Rotating shadow-band radiometer. Rotating band is below silicon photodiode photoelectric pyranometer (just above horizon line).

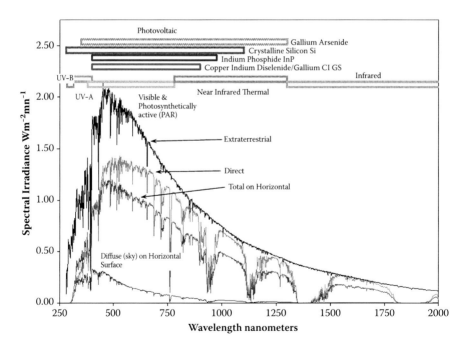

FIGURE 2.6 Representative distributions of power as a function of wavelength for solar energy at the top of the atmosphere (extraterrestrial), solar beam radiation, diffuse sky radiation, and total radiation on a horizontal surface for clear sky conditions. Bars show regions of energy utilization by conversion systems. Note especially the difference between the global and diffuse horizontal spectra.

FIGURE 2.10 Typical absolute cavity pyrheliometer for reference DNI measurements.

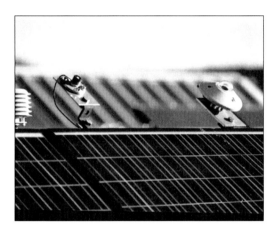

FIGURE 6.1 Global hemispherical pyranometers (left, silicon photodiode; right, thermo-pile) measuring plane of array irradiance for photovoltaic performance testing.

3 Modeling Clear Sky Solar Radiation

Energie is the operation, efflux or activity of any being: as the light of the Sunne is the energie of the Sunne, and every phantasm of the soul is the energie of the soul. [The first recorded definition of the term *energy* in English.]

—Henry More, 1642

3.1 THE ATMOSPHERIC FILTER

Earth's atmosphere is a continuously variable filter operating on the quasi-collimated extraterrestrial direct beam, or direct normal, solar radiation Io, the extraterrestrial radiation (ETR) direct normal irradiance (DNI) as shown in Figures 1.3 and 1.4 of Chapter 1. As discussed in Section 2.4 of Chapter 2 and in Chapter 7, the atmospheric filter modifies not only the intensity of the solar irradiance but also the spectral distribution of the energy. We denote the condition of "clear skies" as the total absence of clouds, which are made up of condensed water vapor droplets, frozen ice crystals, or a mixture of the two. This simplifies the model equations considerably as radiation modeling in the presence of clouds can be daunting. Models accounting for cloud effects are covered in the next chapter.

3.2 PHYSICS-BASED MODELS

The sunlight reaching the top of the atmosphere or ETR is the starting point for models based on the physics of the interaction of matter and radiation, here called *physics-based models*. As described in Chapter 1 on fundamentals, the ETR DNI is a function of the day of the year since the elliptical orbit of Earth brings it closer (perihelion) than the mean Earth–Sun distance in the Northern Hemisphere winter and father away (aphelion) from the mean distance in the Northern Hemisphere summer. The inverse square law ($1/r^2$) dependence of irradiance from a point source with respect to distance r from the source results in 3% greater ETR irradiance in January than the irradiance at the mean Earth–Sun distance (1 AU) and 3% less than the 1-AU ETR irradiance in summer. Physics-based models rely in some degree on the physics of the interaction of the ETR irradiance and the constituents of the atmosphere (discussed in more detail in Chapter 7 on modeling spectral distributions).

3.3 EMPIRICAL MODELS

Empirical models, on the other hand, are based on correlations, or relations (functions) derived, usually through linear or multilinear regression analysis (curve fitting). This approach assumes that measured solar radiation data can be described as a function of some other independently measured or available variables or parameters. These independent variables can range from simple, single variables such as temperature, to more complex combinations of temperature, relative humidity, day length, cloud cover, sunshine duration in hours, and so on.

One of the simplest empirical models is to lump all of the effects of atmospheric transmittance into the parameter of air mass, or path length through the atmosphere m, and a "bulk" attenuation factor. Such a model for computing the direct beam irradiance (DNI) at the surface on a clear day was presented by Meinel and Meinel [1] in 1976. That model requires only the calculation of the extraterrestrial direct beam radiation, 1366.1 Wm^{-2}, multiplied by Rc, the Earth–Sun distance correction (Equation 1.1) and the air mass m (at sea level):

$$DNI = 0.7Rc\ Io^{(0.678\ m)} \tag{3.1}$$

If one begins to consider the "correction terms" needed to account for the particulars of a site or the state of the atmosphere, additional functional terms are needed. For instance, to account for the altitude above sea level, one can modify the sea level air mass by the ratio of the station pressure to sea level. An explicit model involving the altitude h (in kilometers) was derived by Laue [2] in 1970:

$$DNI = Rc\ Io[0.7(1.0 - 0.14h)\ Rc\ Io^{(0.678\ m)} + 0.14h] \tag{3.2}$$

These extremely simple models provide only a back-of-the-envelope estimate of clear sky DNI throughout the day as the air mass decreases and increases. A popular method of estimating the actual attenuation properties of the atmosphere is the Linke turbidity factor T_L proposed by Linke in 1922 [3]. Linke's proposal was based on an optical thickness of clean (aerosol free) and dry (no water vapor) atmosphere τ_D, so that the transmitted DNI irradiance is expressed as

$$DNI = Rc\ Io\ e^{(-\tau_D\ T_L m)} \tag{3.3}$$

with the clean dry atmosphere absorption factor:

$$\tau_D = 0.128 - 0.054\ \log(m). \tag{3.4}$$

For total global hemispherical irradiance (GHI), the usual approach is to estimate the additional contribution of the diffuse horizontal irradiance (DHI) and use the component balance equation, Equation 1.20 of Section 1.4, to compute the GHI from DNI and DHI.

An extensive discussion of empirical models for modeling GHI and DHI radiation and an evaluation of their performance is available in a 2012 technical report by Reno, Hansen, and Stein [4]. They showed the accuracy of these models was on the order of 10%, at best. A more extensive discussion of empirical or correlation models is covered in the next chapter, on modeling solar irradiance under all sky conditions. From here, we discuss more accurate (by at least a factor of 2) parameterization models for clear sky solar radiation that use a combination of correlations and physical principles.

3.4   PARAMETERIZATION MODELS

The primary ingredients of the atmosphere are atmospheric gas molecules, which have finite dimensions on the order of various wavelength light in the ETR DNI spectral power distribution. Various parts of the solar spectrum are either absorbed by or scattered from these molecules. Famously, Lord Rayleigh showed that the sky is blue because the average size of the molecules is such that shorter-wavelength photons (corresponding to the blue region of the human color perception range) are preferentially scattered out of the DNI beam. Thus, these molecules perform "Rayleigh scattering." The mixed gases (nitrogen, oxygen, carbon dioxide, etc.), water vapor, stratospheric ozone, and aerosols, or small scattering centers suspended in the atmosphere, also filter and attenuate the DNI beam as it passes through the atmosphere.

Each of these individual constituents can be considered to have a transmittance (ratio of what impinges at the top of the atmosphere to what remains at the ground level). Each transmittance can be parameterized in terms of the air mass and concentration or amount of a constituent present in the atmosphere. Thus, we consider the total transmittance of the atmosphere T as the product of the terms:

Tr: transmittance due to Rayleigh scattering
Ta: transmittance due to aerosol properties
Tg: transmittance due to optical properties of gases
To: transmittance due to ozone (in the stratosphere)
Tw: transmittance of water vapor

We can write

$$DNI = Io\ Rc\ Tr\ Ta\ Tg\ To\ Tw \tag{3.5}$$

for the direct beam Ib at the observer, derived from the extraterrestrial direct beam irradiance Io modified for the Earth–Sun distance.

This is the formulation of the Bird and Hulstrom clear sky model [5,6], based on previous work by researchers such as Leckner [7], Erbs et al. [8], Watt [9], and Atwater and Ball [10]. There are other formulations recently developed by Gueymard [11] and many others. Next, we outline the computations for three approaches to "physical" models based on correlations of the transmittances with concentrations of constituents and the path length through the atmosphere to the observer.

3.4.1 Bird Clear Sky Direct Beam Irradiance

The transmittance equations derived by Bird and Hulstrom to compute the direct beam component that results from Equation 3.1 are [here exp(a) is the exponential function e^a, and we use M for air mass and Mp for air mass corrected for station pressure]:

Tr = Rayleigh transmittance:

$$Tr = exp(1.0 + Mp - Mp^{1.01})(-0.0903\ Mp^{0.84}) \qquad (3.6)$$

Tg = Mixed gas transmittance:

$$Tg = exp(-0.0127\ Mp^{0.26}) \qquad (3.7)$$

To = Ozone transmittance, for total column ozone amount Oz (in at-cm):

$$Ozm = Oz\ M \qquad (3.8)$$

$$To = 1 - 0.1611Ozm\ (1.0 + 139.48Ozm)^{(-0.3035)}$$
$$- (0.002715Ozm)/(1 + 0.044Ozm + 0.0003Ozm^2) \qquad (3.9)$$

Tw = Water vapor transmittance for precipitable water vapor amount PW (atm-cm):

$$W = PW\ M\ and \qquad (3.10)$$

$$Tw = 1 - 2.4959W/[(1 + 79.034W)^{0.6828} + 6.385W] \qquad (3.11)$$

Ta = Aerosol transmittance:

$$Tau = 0.2758Ta3 + 0.35Ta5 \qquad (3.12)$$

Ta3 = Aerosol optical depth at 380 nm

Ta5 = Aerosol optical depth at 500 nm

$$Ta = exp[(-Tau^{0.873})(1. + Tau - (Tau^{0.7088}))\ M^{0.9108}] \qquad (3.13)$$

Note: "Tau" is based on a rural aerosol distribution of Shettle and Fenn [12], which could be approximated from "routine" measurements from the National Oceanic and Atmospheric Administration (NOAA) at the time of model development. For the present model, it is recommended that available measured aerosol optical depth (AOD) data from the National Aeronautics and Space Administration (NASA) AERONET (Aerosol Robotic Network) network described in the following be used.

So-called broadband aerosol optical depth (BBAOD) may be derived from the attenuation of the direct beam once the attenuation due to ozone and water vapor is accounted for. Molineaux and Ineichen [13] have shown that such BBAOD is equivalent to the spectral aerosol optical depth at a specific wavelength, namely, 700 nm. Since optical depth of an aerosol in a clear sky is generally a decreasing function of increasing wavelength, multiplying a BBAOD by about 1.5 would be equivalent to spectral optical depth at 500 nm, and 1.8 × BBAOD is approximately the spectral AOD at 380 nm. Total precipitable water (the equivalent depth of water in centimeters if condensed out of the entire atmosphere above a location) may be estimated from relative humidity as described by Garrison and Adler [14] using the following equations:

Saturated vapor pressure Es of water vapor in terms of temperature T in kelvin (°C + 271.73°):

$$\text{Log}_{10} \text{ Es} = -8.42926609 - 1827.17833/T - 71208.271/T^2 \text{ (millibar, mB)} \quad (3.14)$$

Pressure-corrected water vapor pressure E from relative humidity RH and station pressure P with respect to sea-level atmospheric pressure of 1013.25 mB:

$$E = RH \text{ Es}(P/1013.25) \quad (3.15)$$

and the estimated water vapor amount W in atmosphere-millimeters (atm-mm):

$$W = 1.45E + 1.5 \text{ (atm-mm)}. \quad (3.16)$$

One source of aerosol and water vapor data needed for the equations that follow is climatological aerosol and water vapor data processed by NASA's AERONET [15], located on the World Wide Web (http://aeronet.gsfc.nasa.gov, accessed 20 July 2012). A typical climatological data table for an AERONET is shown in Table 3.1. Meanings of the columns are as follows:

tau_{a500}: Aerosol optical depth at 500 nm
$alpha_{440-870}$: Angstrom exponent alpha (α), the value of which is related to the particle size distribution (1.3 is a "typical" value often used, but not required, by Bird)
PW: Total column precipitable water vapor, units of atmosphere-centimeters (atm-cm)
N: The number of clear hour values within the months indicated
Month: The number of months over which data were measured
Sigma: The standard deviation of the mean values for the N samples

For ozone, there are some values available on the Internet for total ozone, such as NASA's Total Ozone Monitoring Spectrometer (TOMS) site (http://toms.gsfc.nasa.gov/aerosols/aerosols_v8.html, accessed 21 July 2012). However, there is a popular model due to van Heuklon [16] used by many. This model estimates total column ozone based on the day of the year and the latitude and longitude of the site.

TABLE 3.1

Climatic Atmospheric Parameters for Barcelona, Spain, from NASA AERONET DATA

Overall Averages of	Tau_{a500}	Sigma	$Alpha_{440\text{-}870}$	Sigma	PW	Sigma	N	Month
Jan	0.12	0.08	1.51	0.40	0.99	0.38	121	6
Feb	0.18	0.14	1.51	0.35	1.02	0.34	121	6
Mar	0.19	0.15	1.35	0.42	1.10	0.42	133	6
Apr	0.21	0.11	1.30	0.33	1.48	0.36	130	6
May	0.21	0.10	1.25	0.37	1.84	0.38	136	6
Jun	0.24	0.10	1.36	0.34	2.40	0.44	126	5
Jul	0.24	0.09	1.28	0.34	2.68	0.40	135	5
Aug	0.22	0.09	1.27	0.30	2.72	0.47	117	5
Sep	0.23	0.12	1.31	0.30	2.31	0.58	111	5
Oct	0.20	0.12	1.20	0.36	1.95	0.63	108	5
Nov	0.13	0.09	1.30	0.38	1.26	0.45	93	6
Dec	0.11	0.07	1.51	0.42	1.02	0.40	105	6
Year	0.19	0.05	1.35	0.11	1.73	0.67	1436	67

Source: http://aeronet.gsfc.nasa.gov/new_web/V2/climo_new/Barcelona_500.html.

Note: Here, the latitude φ and longitude Ψ are in degrees, and dn is the day of the year; January 1 = 1.

For the Northern Hemisphere ($\phi > 0$):

If $\Psi s \geq 0$ (Eastern Hemisphere),

$$O3 = 0.001(235.0 + (150.0 + 40.0 \sin(0.017218(dn - 30.0))$$
$$+ 20.0 \sin(0.05256(\Psi + 20.0))) (\sin (0.02234\varphi))^2) \tag{3.17}$$

If $\Psi < 0$ (Western Hemisphere),

$$O3 = 0.001(235.0 + (150.0 + 40.0 \sin(0.017218(dn) - 30.0))$$
$$+ 20.0 \sin (0.017218(dn))(\sin(0.02234\varphi))^2) \tag{3.18}$$

For the Southern Hemisphere ($\varphi \leq 0$) and for $\Psi \geq 0$ (Eastern Hemisphere) and $\Psi < 0$ (Western Hemisphere):

$$O3 = 0.001(235.0 + (100.0 + 30.0 \sin(0.017218(dn + 152.625))$$
$$+ 20.0 \sin(0.03491(\Psi - 75)))(\sin(0.02618\varphi))^2) \tag{3.19}$$

At this point, Equation 3.1 can be computed to give the estimated direct beam at the surface for a particular date, time, and location. What about the diffuse sky radiation and the total hemispherical radiation?

3.4.2 BIRD CLEAR SKY DIFFUSE IRRADIANCE

Diffuse sky radiation is that radiation that has been scattered out of the direct beam and possibly absorbed and reemitted by the constituents in the atmosphere. The photons or "rays" of diffuse radiation are generally scattered in random directions, so are "uncollimated." The scattering, absorption, and reemission processes are complex but are studied in detail by atmospheric physicists. For the purposes of solar energy conversion systems, the approximations for these scattering functions are highly simplified.

For instance, Bird computed the sky scattered radiation on a horizontal surface Ihs and the sky diffuse scattered radiation Ds from

$$Ds = 0.79Io \cos(0.01745\ z)\ To\ Tm\ Tw\ TAA \qquad (3.20)$$

where z is the zenith angle,

$$TAA = 1 - 0.1(1 - M + M^{1.06})(1 - Ta) \qquad (3.21)$$

and

$$Ihs = Ds(0.5(1 - Tr) + 0.85(1 - Ta/TAA))/(1 - M + M^{1.02}) \qquad (3.22)$$

with Tr, To, Tm, Tw, and Ta from Equations 3.6 through 3.13 and the air mass M.

The ground is assumed to have a uniform reflectance, or albedo, generally about 0.2 and up to 0.9 for snow cover. A fraction of the radiation that is reflected upward off the ground is reflected back down to the ground by the atmosphere. Bird determined an expression for this reflected sky radiation Rs based on a forward-scattering ratio Ba = 0.85 and an expression relating the aerosol transmittance TA (Equation 3.13) and aerosol absorptance TAA (Equation 3.21), namely,

$$Rs = 0.0685 + (1 - Ba)(1.0 - TA/TAA). \qquad (3.23)$$

This reflected radiation is added into the diffuse irradiance to complete the computation of the total hemispherical, or global horizontal irradiance (GHI), as described in the following discussion.

3.4.3 THE BIRD CLEAR SKY TOTAL HEMISPHERICAL IRRADIANCE

The clear sky total hemispherical irradiance on a horizontal surface GHI is just the combination of the computed direct beam and the computed sky diffuse irradiances:

Total hemispherical solar radiation GHI for albedo Ab becomes

$$GHI = (Ib \cos(z) + Ias)/(1 - (Ab\ Rs)) \qquad (3.24)$$

And the total diffuse irradiance on a horizontal surface DHI becomes

$$DHI = GHI - DNI \cos(z) \qquad (3.25)$$

3.4.4 COMPUTATIONAL EXAMPLE

Use the previous equations as a recipe for calculating the Bird clear sky model for a particular time and place:

Place: Barcelona, Spain; local albedo = 0.20
Latitude: $\varphi = 41°\ 23"\ N\ (41.38°\ N)$
Longitude: $\Psi = 2°\ 07"\ E\ (2.117°\ E)$
Date: June 15, 2011 (day of the year dn = 166)
Time: 10 a.m. local standard time

Note: The site is in the Greenwich mean time (GMT) time zone 0 (–7.5° < GMT < 7.5° longitude) for solar time calculations.

Site altitude above sea level: 125 m
Mean atmospheric air pressure (from altitude), Equation 1.16:

$$Ps = 989.7\ mB$$

Solar geometry (from Chapter 1, introduction equations (1.1–1.5):
  Earth radius vector correction: T = 2.8403 radians = 2.8403*57.2958°/radians = 162.74°
  Rc = 0.9684
  Extraterrestrial direct beam irradiance: Io = 1322.9 Wm^2
  Solar declination (radians) $\delta = 0.4064$
  Convert to degrees: (0.4064)(57.2958) = 23.2°
  Equation of time (minutes): $E_t = -0.422$ (minutes)

Longitude correction Lc from time zone reference meridian (minutes):

$$Lc = 4*(0.0 - (+2.116) = -8.46\ min$$

True solar time (converting minutes to hours for equation of time and longitude correction):

$$TST = 9.852\ h\ (9:51\ a.m.)$$

Solar hour angle: $\omega = 32.22°$
Geometric solar zenith angle: Z = 0.5663 radians = 32.448°
Air mass: M = 1.1850
Refraction-corrected air mass: Mr = 1.1855

From the refraction-corrected air mass Mr, the refracted zenith angle Zr may be calculated:

Cos(Zr) = 1/Mr = 0.8435, Zr = Acos(0.8435) = 0.56708 radians = 32.49°

Atmospheric parameters:
Ozone (from Heuklon; Equation 2.17): = 0.235 atm-cm
Aerosol optical depth at 500 nm (from AERONET, Table 3.1, annual mean = 0.19, June mean = 0.24)
Water vapor (from AERONET, annual mean = 1.73 atm-cm, June mean = 2.4 atm-cm)

In the following, we use the June monthly average values.

Atmospheric transmittances:
Tr = 0.92 from (3.6)
Tg = 0.98 from (3.7)
To = 0.98 from (3.9)
Tw = 0.88 from (3.10) and (3.11)
Ta = 0.81 from (3.12) and (3.13)

So, DNI = (1322.9)(0.92)(0.98)(0.98)(0.88)(0.81) = 833 Wm^2 direct beam irradiance.
From Equations (3.20), (3.21), and (3.22):

Diffuse scattered radiation Ihs = 137.5 Wm^2

For an albedo Ab of 0.2 and using the default forward-scattering ratio of Ba = 0.85, the reflected sky radiation Rs is

Rs = 0.09 Wm^2

from which the total hemispherical irradiance is

GHI = (833 cos(32.49°) + 137.5)/[(1 − 0.090)(0.2)] = 842 Wm^2

and the diffuse sky radiation is

DHI = 842 − 833 cos(32.49) = 139 Wm^2

3.4.5 THE INEICHEN SIMPLIFIED SOLIS MODEL

Now that we are familiar with the concepts of atmospheric constituents and their transmittance functions, a very simple model developed by Ineichen [17] is described. This is the simplified SOLIS model. The complex SOLIS model described by Mueller [18] was developed as part of the Heliosat-3 European Meteosat satellite irradiance modeling project. It is a spectral physical model based on radiative transfer model calculations. This complex model needs atmospheric water vapor column and aerosol content input parameters. These two parameters can be retrieved from the Aerosol Robotic Network (http://aeronet.gsfc.nasa.gov/new_web/aerosols.html), European Soda databank (http://www.soda-is.com/eng/index.html), or Meteonorm program (http://www.meteonorm.com).

The Ineichen simplified model is based on the typical exponential extinction of the direct beam irradiance:

$$DNI = Io \exp(-\tau m) \tag{3.26}$$

with the usual air mass m, total optical depth of the atmosphere tb, and extraterrestrial (beam) irradiance Io. To account for differing spectral transmission in different wavelength bands, modification of the air mass term is made by an exponent, b. The parameter b was determined by regression analysis of DNI as a function of b from many complex SOLIS model runs, using the sine of the elevation angle e in place of m:

$$DNI = Io \exp(-tb/\sin^b(e)) \tag{3.27}$$

For the GHI, modification of the direct beam by the sine of the solar elevation provides a first estimate of the GHI, and a GHI optical depth tg is used:

$$GHI = Io \exp(-tg/\sin^g(e)) \sin(e) \tag{3.28}$$

Note the exponent g on the sin(e) term is different from the b in the DNI equation.

Similarly, for the diffuse irradiance:

$$DHI = Io \exp(-td/\sin^d(e)) \tag{3.29}$$

with the exponent d for the sin(e) term and diffuse optical depth td.

To address possible high aerosol or excessive scattering situations, a modified extraterrestrial parameter Io′ replaces Io in these equations and was derived using the same regression techniques.

Radiative transfer model calculations were done over a wide range of altitudes and atmospheric conditions to obtain the best fit for the Io′, b, g, d, tb, tg, and td parameters.

The results of the regression analysis for elevations from sea level to 7000 m, water vapor from 0.2 to 10 atm-cm, and AOD (at 700 nm, or equivalent to BBAOD) from 0 to 0.45 were [17]

$$Io' = 1618$$

$$tg = 0.464$$

$$g = 0.402$$

$$tb = 0.606$$

$$b = 0.491$$

$$td = 2.698$$

$$d = 0.187$$

Now, only these parameters and the solar elevation angle are needed to estimate clear sky irradiances using Equations 3.27 to 3.29.

For the Barcelona, Spain, example used in the Bird model in the preceding section, the solar zenith angle was 32.448°, so the elevation angle $e = 90° - 32.44° = 57.55°$, $\sin(e) = 0.8439$, and

$$DNI = 1618 \exp(-0.606/(0.8439)^{0.491}) = 837.4 \text{ Wm}^2 \text{ (Bird: 833 Wm}^{-2})$$

$$GHI = 1618 \exp(-0.464/(0.8439)^{0.402}) \, 0.8439 = 830.8 \text{ Wm}^2 \text{ (Bird: 842 Wm}^{-2})$$

$$DHI = 1618 \exp(-2.698/(0.8430)^{0.187}) = 99.8 \text{ Wm}^2 \text{ (Bird: 137 Wm}^{-2})$$

Note that the model also produces daily irradiance profiles that are symmetrical about solar noon, when the solar elevation angle is at maximum.

DHI irradiances differ by almost 30%. This is because as each component is computed independently of the other, this model has no coupling between the irradiance components. However, if we compute the DHI using the DNI and GHI in the solar component relation,

$$DHI = GHI - DNI \sin(e) = 124.2 \text{ Wm}^2$$

The DHI Bird and simple SOLIS DHI values are within 10% each of other. For the DNI and GHI values, the Bird and simple SOLIS models agree to about 1% or better. This is probably fortuitous in that the site is near sea level and has a relatively clean, dry atmosphere. For sites with largely different water vapor and aerosol climates, Ineichen provided parameterizations of the coefficients based on the AOD at 700 nm and the water vapor. These parameterizations are discussed in the next section.

3.4.6 EXTENSION OF THE SIMPLE SOLIS MODEL

To account for variable AOD and PW climates, as well as site-specific pressure corrections, measured or estimates of input parameters of the station pressure, atmospheric AOD at 700 nm, and water vapor column may be used to produce modified Io′, tb, tg, td, and b, g, d values using the equations that follow. The ozone content is taken constant at 0.340 atm-cm, and the aerosol profile is assumed to be urban. The SOLIS parameterizations are all made with respect to a spectral optical depth at 700 nm, t_{700}, essentially the same as the broadband optical depth derived from a clean dry atmosphere [11]. If not known or available, one can use a linear combination of optical depths at 500 and 380 nm, t_{500} and t_{380}, respectively, as derived by Bird and Hulstrom [5, 6]:

$$t_{700} = 0.27583t_{380} + 0.35t_{500} \tag{3.30}$$

An alternative (rough) rule of thumb is that due to the decreasing optical depth with wavelength, t_{700} is approximately 85% of t_{500}. This is based on comparing the average two AOD in the AERONET climatological data tables from both high- and low-AOD sites. Thus,

$$t_{700} = 0.85t_{500} \tag{3.31}$$

$$Io' = Io[I2\ t^2{}_{700} + I1\ t_{700} + I0 + 0.071\ \ln(p/1013.25)] \tag{3.32}$$

where $I0 = 1.08\ w^{0.0051}$; $I1 = 0.97\ w^{0.032}$; $I2 = 0.12\ w^{0.56}$ for precipitable water vapor w atm-cm, AOD at 700 nm of t_{700}, and station pressure p mB.

For Equation 3.27,

$$tb = tb1\ t700 + tb0 + tbp\ \ln\ (p/1013.25) \tag{3.33}$$

where

$$tb1 = 1.82 + 0.056\ln(w) + 0.0071\ln^2(w) \tag{3.34}$$

$$tb0 = 0.33 + 0.045\ln(w) + 0.0096\ln^2(w) \tag{3.35}$$

$$tbp = 0.0089w + 0.13 \tag{3.36}$$

$$b = (0.00925t^2{}_{700} + 0.0148t_{700} - 0.0172)\ln(w) + b0 \tag{3.37}$$

and

$$b0 = -0.7565t^2{}_{700} + 0.5057t_{700} + 0.4557 \tag{3.38}$$

For Equation 3.28,

$$tg = tg1\ t_{700} + tg0 + tgp\ \ln(p/1013.25) \tag{3.39}$$

where

$$tg1 = 1.24 + 0.047\ln(w) + 0.0061\ln^2(w) \tag{3.40}$$

$$tg0 = 0.27 + 0.043\ln(w) + 0.0090\ln^2(w) \tag{3.41}$$

$$tgp = 0.0079w + 0.10 \tag{3.42}$$

$$g = -0.3079t^2{}_{700} + 0.2846t_{700} - 0.0147\ln(w) + 0.3798 \tag{3.43}$$

and for Equation 3.29,

$$td = td4\ t^4{}_{700} + td3\ t^3{}_{700} + td2\ t^2{}_{700} + td1\ t_{700} + td0 + tdp\ \ln(p/1013.25) \tag{3.44}$$

where td0, … , td4, and tdp have been fit for $t_{700} < 0.05$ and ≥ 0.05, respectively.

For $t_{700} < 0.05$,

$$td4 = 8.06w - 13,800.0 \tag{3.45}$$

$$td3 = -3.11w + 79.4 \qquad (3.46)$$

$$td2 = -0.23w + 74.8 \qquad (3.47)$$

$$td1 = 0.092w - 8.86 \qquad (3.48)$$

$$td0 = 0.0042w + 3.12 \qquad (3.49)$$

$$tdp = -0.83(1.0 - t_{700})^{-17.2} \qquad (3.50)$$

for $t_{700} \geq 0.05$,

$$td4 = -0.21w + 11.6 \qquad (3.51)$$

$$td3 = 0.27w - 20.7 \qquad (3.52)$$

$$td2 = -0.134w + 15.5 \qquad (3.53)$$

$$td1 = 0.0554w - 5.71 \qquad (3.54)$$

$$td0 = 0.0057w + 2.94 \qquad (3.55)$$

$$tdp = -0.71(1.0 - t_{700})^{-15.0} \qquad (3.56)$$

and exponent d,

$$d = -0.337t^2{}_{700} + 0.63t_{700} + 0.116 + [1.0/(18.0 + 152.0t_{700})] \ln(p/1013.25) \qquad (3.57)$$

We can now recompute our Barcelona example in Section 3.1 using $t_{500} = 0.24$ ($t_{700} = 0.20$), w = 2.4 atm-cm, 989.7-mB station pressure; so $\ln(p/1013.25) = -0.0235$. Substituting in Equation 3.29 (using Io = 1366 wm^{-2}): Io′ = 1762.8

From Equations 3.33 to 3.38: tb = 0.7481, b = 0.5144.

From Equations 3.39 to 3.43: tg = 0.5696, g = 0.4362.

From Equations 3.44 and 3.51 to 3.57: td = 2.6688, d = 0.2280.

And, entering these values into Equations 3.27 to 3.29 for the given elevation angle of 57.55°,

DNI = 1762 exp(−0.7481/(0.8439)$^{0.514}$) = 779.2 Wm^2 (Bird: 833 Wm^{-2})

GHI = 1762 exp(−0.5690/(0.8439)$^{0.436}$) 0.8439 = 805.6 Wm^2 (Bird: 842 Wm^{-2})

DHI = 1762 exp(−2.6669/(0.8430)$^{0.228}$) = 110.0 Wm^2 (Bird: 137 Wm^{-2})

Now, the differences between Bird and the extended simple SOLIS model are somewhat larger for the DNI (6.5%) and GHI (4.4%) but 30% less for the DHI (19.7%). But, computing the least-accurate model value (DHI) from modeled DNI and GHI,

$$DHI = 805.6 - (0.8439)*779.2 = 148.0 \ Wm^{-2},$$

within 10% of the Bird DHI model.

At least in this example, the additional information from the t500 and w data seems to improve agreement between the Bird and extended simple SOLIS models in the diffuse. The differences between the BIRD and the extended model values are larger for the DNI and GHI, but only on the order of 30 to 40 wm^{-2}. This single example again may only be fortuitous; careful comparison of models with the same common input parameters over a wide range of values is the best way to evaluate and compare models. It is also clear that, depending on the climatology for w and AOD at the site, these model differences will no doubt be different.

We now move to a more complex empirical model, expanding on the principles of the Bird model, developed by Gueymard.

3.4.7   GUEYMARD'S REST2 MODEL

REST2 is a state-of-the-art, high-performance model to predict cloudless sky broadband irradiance, illuminance, and photosynthetically active radiation (PAR) from atmospheric data. The model was developed by Dr. C. Gueymard [11]. It is based on a two-band scheme. The solar irradiance is broken down into two spectral regions:

Region 1: 290–700 nm (about 46.5% of the total energy) and
Region 2: 700–4000 nm (about 52% of the total)

The first region is affected mainly by aerosols and Rayleigh scattering, the second by water vapor and gaseous absorption. Note the sum of the percentages is 98.5% of the "total" irradiance. This is because the model is based on parameterizations of his spectral model SMARTS2 [19], which only compute the solar spectrum from 290 to 4000 nm. There is about 1.5% to 2.0% of the total solar irradiance beyond 4000 nm. The slight underestimate in the result is a small bias that can be added into the model result. However, this difference is smaller than the uncertainty in the best well-maintained solar radiation measurement instrumentation (about 5%) (see references 20–23). This effect is discussed in the validation against measured data presented in the article describing the model. A package, containing an executable, examples, and necessary information, is available for the Mac OS X platform with Intel processors and for PC Windows from the Solar Consulting Services Web site (http://www.solar-consultingservices.com/REST2_model.php, accessed 20 July 2012).

The REST2 model uses the following, by now familiar, inputs:

• Site coordinates (latitude, longitude, elevation, time zone)
• Station pressure (calculated from site elevation and latitude if unknown)

- Precipitable water (estimated from temperature and relative humidity if unknown)
- Angstrom turbidity coefficient
- Angstrom wavelength exponent
- Aerosol single-scattering albedo (defaulted if unknown)
- Total columnar ozone amount
- Total columnar nitrogen dioxide amount (defaulted if unknown)
- Ground albedo

3.4.7.1   Basic REST2 Structure

The REST2 equations are similar to the Bird model equations but address additional gases (nitrogen dioxide, NO_2) and different functions for transmittances in each of regions i = 1 and 2 in a more complex and complete fashion. The model equations take the following form:

$$Ebn_i = Tr_i\ Tg_i\ To_i\ Tn_i\ Tw_i\ Ta_i \qquad (3.58)$$

with i = 1 and 2, and the total beam irradiance the sum of Ebn_1 and Ebn_2 is

$$DNI = Ebn_1 + Ebn_2 \qquad (3.59)$$

Diffuse hemispherical irradiance Edf_i in each of the two bands i = 1 and 2 are from

$$Edf_i = To_i\ Tg_i\ Tn_i\ Tw_i[Br_i\ (1 - Tr_i)\ Ta_i^{0.25} + Ba\ F_i\ Tr_i(1 - Ta_i^{0.25})]\ Io_i \quad (3.60)$$

where $Io_i = Ion_i$, and Ion_i is the extraterrestrial direct beam irradiance in band i corrected for the earth orbit radius vector length.

$$Br1 = 0.5(0.89013 - 0.0049558mr + 0.000045721mr^2) \qquad (3.61)$$

$$Br2 = 0.5 \qquad (3.62)$$

are the forward-scattering fractions for Rayleigh extinction in each band, but negligible in band 2. Rather than a single air mass m to characterize the solar path length through the atmosphere, individual optical masses mr, mo, mw, and ma, computed independently, are used for Rayleigh (molecular) scattering and uniformly mixed gases absorption, ozone absorption, water vapor absorption, and aerosol extinction, respectively.

3.4.7.2   The REST2 Model Transmittance Equations

These relevant constituent specific air masses and the transmittance parameters in Equations (3.59–3.62 are defined in the REST2 model transmittance equations (air mass terms with primes are pressure corrected for station pressure):

Transmittance band 1:

$$Tr1 = \text{Rayleigh scattering} = (1 + 1.8169 \, mg' - 0.033454 \, mr'^2)/$$
$$(1 + 2.063 \, mr' + 0.31978 \, mg'^2) \tag{3.63}$$

Tg_1 = Mixed gas absorption dependent on Rayliegh air mass mr'

$$= (1 + 0.95885 mr' + 0.012871 mr'^2)/(1 + 0.096321 mr' + 0.015455 mr'^2) \tag{3.64}$$

$$mr' = mr \, p/1013.25 \tag{3.65}$$

$To1$ = Ozone absorption, for uo (atm-cm):

$$To1 = (1 + f1 \, mo + f2 \, mo^2)/(1 + f3 \, mo) \tag{3.66}$$

where

$$f1 = uo(10.979 - 8.5421uo)/(1 + 2.0115uo + 40.189uo^2) \tag{3.67}$$

$$f2 = uo(0.027589 - 0.005138uo)/(1 - 2.4857uo + 13.942uo^2) \tag{3.68}$$

$$f3 = uo(10.995 - 5.5001uo)/(1 + 1.6784uo + 42.406uo^2) \tag{3.69}$$

$Tn1 = NO_2$ absorption, for un (atm-cm)

$$= \text{Min} [1, 1 + g1 \, mw + g2 \, mw^2)/(1 + g3 \, mw) \tag{3.70}$$

$$g1 = (0.17499 + 41.654un - 2146.4un^2)/(1 + 22295.0un^2) \tag{3.71}$$

$$g2 = un(-1.2134 + 59.324un)/(1 + 8847.8un^2) \tag{3.72}$$

$$g3 = (0.17499 + 61.658un + 9196.4un^2)/(1 + 74109.0un^2) \tag{3.73}$$

$Tw1$ = Water vapor absorption, for w (atm-cm)

$$= (1 + h1 \, mw)/(1 + h2 \, mw) \tag{3.74}$$

$$h1 = w(0.065445 + 0.00029901w)/(1 + 1.2728w) \tag{3.75}$$

$$h2 = w(0:065687 + 0.0013218w)/(1 + 1.2008w) \tag{3.76}$$

Transmittance band 2:

$$Tr2 = (1 - 0.010394mr')/(1 - 0.00011042mr'^2) \tag{3.77}$$

$$Tg2 = (1 + 0.27284mr' - 0.00063699mr'^2)/(1 + 0.30306mr') \tag{3.78}$$

$$To2 = 1 \tag{3.79}$$

$$Tn2 = 1 \tag{3.80}$$

$$Tw2 = (1 + c1\ mw + c2\ mw^2)/(1 + c3\ mw + c4\ mw^2) \tag{3.81}$$

$$c1 = w\ (19.566 - 1.506w + 1{:}0672w^2)/(1 + 5.4248w + 1.6005w^2) \tag{3.82}$$

$$c2 = w\ (0.50158 - 0.14732w + 0.047584w^2)/(1 + 1.1811w + 1.0699w^2) \tag{3.83}$$

$$c3 = w(21.286 - 0.39232w + 1.2692w^2)/(1 + 4.8318w + 1.412w^2) \tag{3.84}$$

$$c4 = w(0.70992 - 0.23155w + 0.096514w^2)/(1 + 0.44907w + 0.75425w^2) \tag{3.85}$$

Effective aerosol wavelength, band 1:

$$\lambda e1 = (do + d1\ ua + d2\ ua^2)/(1 + d3\ ua^2) \tag{3.86}$$

$$d0 = 0.57664 - 0.024743\alpha1 \tag{3.87}$$

$$d1 = (0.093942 - 0.2269\alpha1 + 0.12848\alpha1^2)/(1 + 0.6418\alpha1) \tag{3.88}$$

$$d2 = (-0.093819 + 0.36668\alpha1 - 0.12775\alpha1^2)/(1 - 0.11651\alpha1) \tag{3.89}$$

$$d3 = \alpha1\ (0.15232 - 0.087214\alpha1 + 0.012664\alpha1^2)/$$
$$(1 - 0.9045\alpha1 + 0.26167\alpha1^2) \tag{3.90}$$

Effective aerosol wavelength, band 2:

$$\lambda e2 = (e0 + e1\ ua + e2\ ua^2)/(1 + e3\ ua) \tag{3.91}$$

$$e0 = (1.183 - 0.022989\alpha2 + 0.020829\alpha2^2)/(1 + 0.11133\alpha2) \tag{3.92}$$

$$e1 = (-0.50003 - 0.18329\alpha2 + 0.23835\alpha2^2)/(1 + 1.6756\alpha2) \tag{3.93}$$

$$e2 = (-0.50001 - 1.1414\ \alpha2 + 0.0083589\alpha2^2)/(1 + 11.1682\alpha2) \tag{3.94}$$

$$e3 = (-0.70003 - 0.73587\alpha2 + 0.51509\alpha2^2)/(1 + 4.7665\alpha2) \tag{3.95}$$

In Equations 3.87 to 3.91 and 3.93 to 3.95 the alphas $\alpha1$ and $\alpha2$ are user input (or default) aerosol parameters for wavelengths below and above 700 nm, respectively.

Aerosol scattering correction factor, band 1:

$$F1 = (g0 + g1\ \tau a1)/(1 + g2\ \tau a1) \tag{3.96}$$

$$g0 = (3.715 + 0.368ma + 0.036294ma^2)/(1 + 0.0009391ma^2) \tag{3.97}$$

$$g1 = (-0.164 - 0.72567ma + 0.20701ma^2)/(1 + 0.0019012ma^2) \tag{3.98}$$

$$g2 = (-0.05228 + 0.31902ma + 0.178171ma^2)/(1 + 0.0069592ma^2) \quad (3.99)$$

Aerosol scattering correction factor, band 2:

$$F2 = (h0 + h1\ \tau a2)/(1 + h2\ \tau a2) \tag{3.100}$$

$$h0 = (3.4352 + 0.65267ma + 0.00034328ma^2)/(1 + 0.034388ma^{1.5}) \quad (3.101)$$

$$h1 = (1.231 - 1.63853ma + 0.20667ma^2)/(1 + 0.1451ma^{1.5}) \quad (3.102)$$

$$h2 = (0.8889 - 0.55063ma + 0.50152ma^2)/(1 + 0.14865ma^{1.5}) \quad (3.103)$$

Sky albedo, for backscattered radiation:

$$\rho s1 = [0.13363 + 0.00077358\alpha1 + \beta1(0.37567 + 0.22946\alpha1)/$$
$$(1 - 0.10832\alpha1)]/[1 + \beta1(0.84057 + 0.68683\alpha1)/(1 - 0.08158\alpha1)] \quad (3.104)$$

$$\rho s21 = [0.010191 + 0.00085547\alpha2 + \beta2(0.14168 + 0.062758\alpha2)/$$
$$(1 - 0.19402\alpha2)]/[1 + \beta2(0.58101 + 0.17426\alpha2)/(1 - 0.17586\alpha2)] \quad (3.105)$$

A detailed validation analysis has shown that the model predictions are within instrumental uncertainty (about 5%) [19] when the inputs are known with good accuracy.

3.4.7.3   REST Computational Example

Rather than show a detailed (and tedious) step-by-step calculation as we did for the Bird model, we run the REST2 model downloaded from Solar Consulting Services (version 5) for the same site as the example for the Bird model in Section 3.4.4:

Place: Barcelona, Spain; local albedo = 0.20
Latitude: ϕ = 41 23 N (41.38° N)
Longitude: Ls = 2 07 E (2.117° E)
Date: June 15, 2011 (month 6, day 15)
Time: 10 a.m. local standard time
GMT (universal time): Time zone 0 = 10.0 h
Site altitude above sea level: 125 m (0.125 km)
Mean atmospheric air pressure (from altitude): Ps = 989.7 mB

The input data required to run the REST2 model are as follows:

Iaer: Data switch to select the proper aerosol climatology
0: Angstrom's alpha1, alpha2 are provided
   [alpha1: wavelength exponent below 0.7 μm]
   [alpha2: wavelength exponent above 0.7 μm]
1: RURAL AREA, VERY CLEAR
2: RURAL AREA, CLEAR

3: URBAN AREA, MODERATELY CLEAR
4: URBAN AREA, POLLUTED
5: URBAN AREA, VERY POLLUTED

Here, we use Iaer = 1 to imply a rural site with relatively clean air, which will default the alphas appropriately.

Idata: Main data switch
 0 if zenith angle is input
 1 if date/time is input with year/month/day/hour (see also Ihour)
Ihour: selects the correct time reference (only used if Idata = 1)
 0 for LST;
 1 for UT/GMT.
Iwater: Data switch for precipitable water
 0 if precipitable water is provided (e.g., if w is known)
 1 if w must be calculated from temperature and relative humidity
Latitude: Location latitude, + for north, – for south (required, but only needed
 if Idata = 1 or if pressure value is missing); use –999.9 if unknown
Longitude: Location longitude, + for east, – for west (required, but only needed
 if Idata = 1); use –999.9 if unknown
Month: integer month number, such as 3 for March (only needed if Idata = 1)
pi01, pi02: Aerosol single-scattering albedo below and above 0.7 μm, respec-
 tively (only needed if Iaer = 0); use –999.9 if unknown
Relative humidity (%) (only needed if Iwater = 1)
Regional ground albedo; use –999.9 if unknown
Air temperature (degrees centigrade) (only needed if Iwater = 1)
NO_2 (nitrous oxide) vertical path length (atm-cm)
Ozone vertical path length (atm-cm); should be less than 0.6 atm-cm. Use
 –999.9 if unknown.
Precipitable water (atm-cm), should be less than 10 cm
Year: Four-digit integer, such as 2003 (only needed if Idata = 1)
Sun's zenith angle (degrees) (only needed if Idata = 0, should be between 0°
 and 90°)
Time zone in hours from Greenwich, + for east, – for west (required, but only
 used if Idata = 1); should be between –12 and +12; use –999.9 if unknown

The input format is a text file of values and flags indicating which parameters are known or estimated and which are unknown (and will have default values automatically used)

The text input file REST2_inp.txt should read as follows:

```
[Card 1] Comment (e.g., 'location_XY'), 96 characters max.,
  no spaces!
[Card 2] Idata, Iaer, Iwater (0 = Z, 1 = date/time for Idata,
  0 = given, 1 = unknown for aerosol/water)
[Card 3—ONLY IF Iaer = 0] Alfa1, Alfa2, Pi01, Pi02c
```

```
[Card 4—ONLY IF Idata = 1] Ihour, Latit, Longit, Altit, Zone
[Card 5] rog
[Card 6a—ONLY IF Idata = 0 and Iwater = 0] Z, p, uo, un, beta, w
[Card 6b—ONLY IF Idata = 0 and Iwater = 1] Z, p, uo, un, beta,
  T, RH
[Card 6c—ONLY IF Idata = 1 and Iwater = 0] Year, Month, Day,
  Hour, p, uo, un, beta, w
[Card 6d—ONLY IF Idata = 1 and Iwater = 1] Year, Month, Day,
  Hour, p, uo, un, beta, T, RH
```

We use Iaer = 1 to imply a rural, relatively clean air site. The AERONET τ at 500 nm is 0.24 for June. The REST2 model expects a value for beta and no turbidity. Our input for beta (β) = 0.1 is based on the fact that the turbidity $\tau(\lambda)$ is computed from $\tau(\lambda) = \beta \lambda^{(-\alpha)}$. Since for the two spectral regions alpha is allowed to default to 1.3, at 500 nm we expect for $\tau(500) = \tau (0.5 \ \upsilon m)$, $\beta = \tau(0.5)/(0.5^{(-1.3)}) = 0.24/2.46 = 0.097$.

Therefore, in this case, our input file reads as follows:

```
[Card 1] Barcelona_Jun_15_10.0_am_clear_sky
[Card 2] 1, 1, 0
[Card 3—ONLY IF Iaer = 0]— NOT USED
[Card 4—ONLY IF Idata = 1] 1, 41.38, 2.117, 0.125, 0
[Card 5] 0.2
[Card 6a—ONLY IF Idata = 0 and Iwater = 0] NOT USED
[Card 6b—ONLY IF Idata = 0 and Iwater = 1] NOT USED
[Card 6c—ONLY IF Idata = 1 and Iwater = 0] 2002, 6, 15, 10.0,
  989.7, 0.235, 0.0002, 0.1, 2.4
[Card 6d—ONLY IF Idata = 1 and Iwater = 1] NOT USED
```

Or, explicitly:

```
Barcelona_Jun_15_10.0_am_clear_sky
1, 1, 0
1, 41.38, 2.17, 0.125, 0
0.2
2012, 6, 15, 10.0, 989.7, 0.235, 0.0002, 0.10, 2.4
```

The resultant output file (REST2_out.txt) is

```
        *** REST2 v5 ***
C. Gueymard, Solar Consulting Services, 2006
------------------------------------

Barcelona_Jun_15_10.0_am_clear_sky
Ground albedo = 0.200; Angstrom's alpha = 1.050
Latitude = 41.380; Altitude (km) = 0.125
```

(Note: Each "case" or output record such as what follows is computed for each input line as one line, broken here for readability. Irradiance units are in watts per square meter.)

```
------------------ ATMOSPHERIC CONDITIONS -----------------
Case#   Z     air_mass   p      w      uo    beta  tau550  tauA
 1    29.497   1.148   989.7  2.400  0.235  0.100  0.245  0.167


-------- IRRADIANCE -----     ------ ILLUMINANCE ----      -------- PAR --------
Bnorm  Bhor   Dhor   Ghor       Blum   Dlum   Glum         Bpar    Dpar    Gpar
792.9  690.1  188.2  878.3      73.85  27.56  101.41       283.98  120.23  404.21
```

Note the REST2 program also produces daylighting values of illuminance (in kilolux) and PAR (Wm²) for agricultural and biomass applications (for beam, diffuse, and global total hemispherical components). Compare these results to those for the BIRD model:

Direct beam radiation DNI = 833 Wm² (5% greater than REST = 793 Wm²)

Diffuse sky radiation DHI = 139 Wm² (35% smaller than REST = 188 Wm²)

Total hemispherical radiation GHI = 842 Wm² (4% smaller than REST = 878 Wm²)

3.5   SUMMARY

The models discussed provide examples of the differences that arise from different approaches to clear sky solar radiation modeling. These differences are typical of the range of results from various clear sky models that use (1) different input parameters; (2) different solar geometry and air mass calculations—note the differences in air mass; (3) different parameterizations for the transmittance functions; (4) algorithms derived from different approaches, either data fitting and correlations or theoretical calculations or combinations of the two approaches; and (5) choice of extraterrestrial "solar constant" (1367 for Bird, 1366.1 for REST2, 1618 in the basic SOLIS model).

In most cases, the most important results for energy applications, direct beam and global hemispherical radiation, are within about 5–10% of each other. See the work of Ineichen [21] for an evaluation of the performance of eight different clear sky models with respect to 16 measurement site databases. It is also good to remember the approximate accuracy attributed to instantaneous measurements for typical solar radiation instrumentation, as discussed in Chapter 2 and References 20, 22, and 23.

REFERENCES

1. Meinel, A.B., and M.P. Meinel. (1976). *Applied Solar Energy*. Addison-Wesley.
2. Laue, E.G. (1970). The measurement of solar spectral irradiance at different terrestrial elevations. *Solar Energy*, Vol. 13, pp. 43–50.
3. Linke, F. (1922). Transmissions-Koeffizient und Trubungsfaktor. *Beitraege zur Physik der Atmosphaere,* Vol. 10, pp. 91–103.
4. Reno, M.J., C.W. Hansen, and J.S. Stein. (2012). *Global Horizontal Irradiance Clear Sky Models: Implementation and Analysis*. Sandia Report SAND2012-2389. http://energy.sandia.gov/wp/wp-content/gallery/uploads/SAND2012-2389_ClearSky_final.pdf. Accessed 16 June 2012.
5. Bird, R.E., and R.L. Hulstrom. (1980). *Direct Insolation Models*. Report SERI/TR-335-344. Solar Energy Research Institute, Golden, CO (now National Renewable Energy Laboratory).

6. Bird, R.E., and R.L. Hulstrom. (1981). Review, evaluation, and improvement of direct irradiance models. *Transactions of the ASME, Journal of Solar Energy Engineering*, Vol. 103, pp. 182–192.

7. Leckner, B. (1978). The spectral distribution of solar radiation at the Earth's surface—Elements of a model. *Solar Energy*, Vol. 20, pp. 143–150.

8. Erbs, D.G., S.A. Klein, and J.A. Duffie. (1982). Estimation of the diffuse radiation fraction for hourly, daily, and monthly-average global radiation. *Solar Energy*, Vol. 20, No. 4, pp. 295–302.

9. Watt, D. (1978). *On the Nature and Distribution of Solar Radiation*. U.S. Department of Energy Report HCP/T2552-01. U.S. Department of Energy, Washington, DC.

10. Atwater, M. A., and J.T. Ball. (1978). A numerical solar radiation model based on standard meteorological observations. *Solar Energy*, Vol. 20, pp. 163–170.

11. Gueymard, C.A. (2008). REST2: High-performance solar radiation model for cloudless-sky irradiance, illuminance, and photosynthetically active radiation—Validation with a benchmark dataset. *Solar Energy*, Vol. 82, pp. 272–285.

12. Shettle, E.P., and R.W. Fenn. (1979). *Models for the Aerosols of the Lower Atmosphere and the Effects of Humidity Variations on Their Optical Properties*. Report AFGL-TR-79-0214, Air Force Geophysics Laboratory, Hanscom, MA.

13. Molineaux, B., and P. Ineichen. (1998). Equivalence of pyrheliometric and monochromatic aerosol optical depths at a single key wavelength. *Applied Optics*, Vol. 37, No. 30, pp. 7008–7018.

14. Garrison, J.D., and G.P. Adler. (1990). Estimation of precipitable water over the United States for application to the division of solar radiation into its direct and diffuse components. *Solar Energy*, Vol. 44, No. 4, pp. 225–241.

15. Holben, B.N., D. Tanre, A. Smirnov, T.F. Eck, I. Slutsker, N. Abuhassan, W.W. Newcomb, J. Schafer, B. Chatenet, F. Lavenue, Y.J. Kaufman, J. Vande Castle, A. Setzer, B. Markham, D. Clark, R. Frouin, R. Halthore, A. Karnieli, N.T. O'Neill, C. Pietras, R.T. Pinker, K. Voss, and G. Zibordi. (2001). An emerging ground-based aerosol climatology: Aerosol Optical Depth from AERONET. *Journal of Geophysical Research*, Vol. 106, 12 067–12 097. http://aeronet.gsfc.nasa.gov/new_web/PDF/AERONET_climo.pdf. Accessed 16 June 2012.

16. Van Heuklon, T.K. (1979). Estimating atmospheric ozone for solar radiation models. *Solar Energy*, Vol. 22, pp. 63–68.

17. Ineichen, P. (2008). A broadband simplified version of the Solis clear sky model. *Solar Energy*, Vol. 82, No. 8, 758–762.

18. Mueller, R.W., K.F. Dagestad, P. Ineichen, M. Schroedter, S. Cros, D. Dumortier, R. Kuhlemann, J.A. Olseth, C. Piernavieja, C. Reise, L. Wald, and D. Heinemann, D. (2004). Rethinking satellite based solar irradiance modelling—The SOLIS clear sky module. *Remote Sensing of the Environment*, Vol. 91, pp. 160–174.

19. Gueymard, C. (2001). Parameterized transmittance model for direct beam and circumsolar spectral irradiance. *Solar Energy*, Vol. 71, No. 5, pp. 325–346.

20. Gueymard, C.A., and D.R. Myers. (2009). Evaluation of conventional and high-performance routine solar radiation measurements for improved solar resource, climatological trends, and radiative modeling. *Solar Energy*, Vol. 83, pp. 171–185.

21. Ineichen, P. (2006). Comparison of eight clear sky broadband models against 16 independent data banks. *Solar Energy*, Vol. 80, pp. 468–478.

22. Myers, D.R. (2005). Solar radiation modeling and measurements for renewable energy applications, data and model quality. *Energy*, Vol. 30, pp. 1517–1531.

23. Myers, D.R., and K.A. Emery. (1989). Uncertainty estimates of global solar irradiance measurements used to evaluate PV device performance. *Solar Cells,* Vol. 27, pp. 455–464.

4 Modeling Global Irradiance under All Sky Conditions

Algebra applies to the clouds.

—Victor Hugo, 1862

4.1 SIMPLE CORRELATION MODELS

There is a popular saying that correlation does not imply cause and effect. For instance, over the lifetime of most of us, as our age increased over time the stock market fluctuated (sometimes a great deal). Overall, throughout our lifetimes, stock exchange indexes have also increased. Therefore, there is a positive correlation of stock indexes with age. However, it would be risky to predict short-term income from the stock market, even with a diversified portfolio, as a function of a person's age.

There are some truly physically induced correlations between solar radiation values and meteorological variables, particularly air temperature, changing atmospheric constituents, and observed cloud cover. Another popular correlation is that between maximum theoretically available hours of sunshine and actual hours of sunshine (defined with suitable lower threshold irradiance, representing "no sunshine"). A brief description of radiation estimates using each of these approaches follows.

4.1.1 SOLAR RADIATION FROM TEMPERATURE OBSERVATIONS

It is intuitive that solar radiation and ambient temperature are well correlated. The air temperature is higher when the sun shines than when it does not. The agricultural research community has long studied the relationship between temperature and solar radiation to develop models for evapotranspiration of water to determine irrigation needs. A famous model developed by Hargreaves and Samani [1] relates daily total (summed over all hours of daylight) solar radiation GHI_d to the difference between maximum and minimum daily ambient temperatures Tx and Ti (in °C), respectively. From daily average extraterrestrial radiation Io_d,

$$GHI_d = Ct\ Io_d\ (Tx - Ti)^{0.5} \qquad (4.1)$$

where Hargreaves and Samani [2] suggested the temperature coefficient $Ct = 0.162$ for "interior" locations and 0.19 for coastal regions.

For instance, for a day where the daily total for $Io_d = 10.2$ kWh/m² (total of extra-terrestrial radiation on a horizontal surface for the period of daylight), if $Tx = 28°C$, $Ti = 10°C$ for an inland site, for a difference of 18°C, from Equation 4.1:

$$GHI_d = (0.162)(10.2)(18)^{0.5} = 7.01 \text{ kWh/m}^2.$$

Other authors have pursued investigations of this model. For instance, Knapp et al. [3] developed a relationship for Ct as a function of the temperature difference ($\Delta T = Tx - Ti$):

$$Ct = 0.00185(\Delta T)^2 - 0.0433\Delta T + 0.4023 \qquad (4.2)$$

Allen [4] suggested using

$$Ct = 0.17(P/Po)^{0.5} \qquad (4.3)$$

for interior regions and

$$Ct = 0.2(P/Po)^{0.5} \qquad (4.4)$$

for coastal regions to account for proximity to a large body of water and elevation effects, where P is station atmospheric pressure and Po is sea-level atmospheric pressure.

Allen [4] showed Equation 4.1 has maximum daily total errors on the order of 50%, whereas Equation 4.3 has maximum daily total errors of about 15%.

4.1.2 CORRELATIONS WITH SUNSHINE DURATION

One of the most famous and most exercised models for estimating solar radiation is based on the duration of actual sunshine versus total possible sunshine. The model was developed by Angstrom [5] and is referred to as the Angstrom sunshine model:

$$Gm/Gm(0) = \alpha + (1 - \alpha)S \qquad (4.5)$$

where Gm(0) is monthly average clear sky total irradiance, and S is the monthly average fraction (of possible maximum) sunshine. α is an empirical constant (Angstrom reported 0.25 for data from Stockholm, Sweden).

Later, Prescott [6] replaced the clear sky irradiance term Gm(0) with the more easily computed extraterrestrial available monthly average solar irradiance Ho and re-formed the model as the Angstrom–Prescott model:

$$Gm/Ho = a + b S \qquad (4.6)$$

A comparison of the two approaches by Angstrom [7] showed that α in Equation 4.5 is related to a and b by

TABLE 4.1

Examples of Angstrom–Prescott Model Coefficient Variability

Author	Date	Location	Range of a	Range of b
Glover and McCulloch [11]	1957	South Africa	0.15 to 0.29	0.48 to 0.60
Almorox et al. [12]	2005	Toledo Spain	0.28 to 0.33	0.41 to 0.49
Tymvios et al. [13]	2005	Globally	0.17 to 0.43	0.24 to 0.75

$$\alpha = a/(a + b) \tag{4.7}$$

A comprehensive review of the application and variations on this particular model is beyond the scope of this chapter. However, there are excellent review articles addressing issues with this model in Suehrcke [8] and Gueymard et al. [9]. Basically, the coefficients α, a, and b are site dependent, and no "universal" coefficients exist. The model requires both the sunshine duration and the measured GHI (global horizontal irradiance) to develop the coefficients. The definition of *bright sunshine* is somewhat illusive. In general, the term is used to signify or represent hours of solar irradiance above a certain threshold. The total possible maximum sunshine hours are then represented by the length of daylight at a site (where the threshold is exceeded).

Currently, the World Meteorological Organization [10] set the lower limit of bright sunshine as 120 Wm^2 of direct beam irradiance. Measurement methods for recording hours of bright sunshine have varied widely over time. Last, while the correlations discussed have been for monthly data, similar approaches to modeling daily data total global hemispherical radiation have been attempted.

For illustrative purposes, Table 4.1 shows the range of Angstrom coefficients derived by several authors. From the table, it appears that average values of a and b for rough estimation of monthly total solar radiation from sunshine hour data are approximately a = 0.28 and b = 045; however, the results could have a great deal (greater than 30%) of uncertainty.

4.2 CLOUDS

The largest influence on solar radiation available for solar energy applications at a specific site is the diurnal day–night cycle. The next largest influence is the cloud climatology of the site. We are all familiar with the fact that a wide variety of cloud types, structures, and distributions are possible. Clouds have a three-dimensional characteristic, making the modeling of solar radiation transfer through them and the atmosphere very complicated. Professional radiation transfer model developers look into the basic physics and esoteric optical properties of water vapor, water droplets, and ice crystals, both in isolation and aggregated into clouds (see [14]).

4.2.1 CLOUD OBSERVATIONS

National weather services around the world report a wide variety of cloud observations based on either human or automated observing systems. Basic cloud data range from sky dome coverage to cloud type, layer, altitude, and whether the clouds are translucent or opaque. In the past, cloud and cloud coverage atlases were developed [15–17].

With the advent of the space age, meteorological satellites, and lately satellites devoted to cloud studies from space [18,19], various cloud parameters from simple cloud cover and motion to optical properties of clouds are being studied from above the atmosphere.

The International Satellite Cloud Climatology Program (ISCCP) [19] coordinates and collates data from international cloud-observing satellite programs. The archived and future data collected by this program may serve as a good resource for all future sky solar radiation model development. The present state of the art in application of weather satellite images for modeling solar radiation is discussed briefly in Section 4.3. The details of these models are beyond the scope of this book due to the vast amount of information retrieval, storage, data processing, and computing power required in producing the results.

The complexities described have led solar radiation model developers for energy conversion system applications to develop empirical engineering approaches to modeling how clouds have an impact on the transfer of solar radiation through the atmosphere.

4.3   EMPIRICAL ALL SKY RADIATION MODELS

4.3.1   KASTEN AND CZEPLAK MODELS

Many authors have attempted parameterizations or correlations of measured solar radiation and cloud observations. Examples include Kasten and Czeplak [20], Gul, Muneer, and Kambezidis [21], and recently Chen et al. [22]. A discussion of cloud cover radiation models can be found in Chapter 3, Section 3.3.5, of Muneer's book [23]. There are significant issues with most of these models. They require input data directly related to solar measurements in the first place, such as direct-to-global or direct-to-diffuse ratios, sunshine duration, Linke turbidity factor (bulk direct normal irradiance [DNI] transmittance of the atmosphere), clearness index (ratio of terrestrial to extraterrestrial radiation), and more. The Kasten and Czeplak model is a very good example, presenting some simple model results. That article also contains 24 references to various attempts at these correlation studies. Some studies relied on basic cloud observations (cloud cover in 8ths, 10ths, or percentages). Some used more detailed information (cloud types, cloud layer information). The Kasten and Czeplak article itself analyzed radiation correlations with cloud coverage and type. They determined that global hemispherical irradiance on a horizontal surface (GHI) as a function of cloud amount N (in eighths) is

$$G(N) = G(0)(1 - 0.75(N/8)^{3.4})$$ (4.8)

where G(0) is the clear sky solar GHI. G(0) can of course be modeled from one of the clear sky models for the particular time of the cloud observation. However, Kasten and Czeplak also reported that the clear sky G(0) value at solar elevation angle e can be estimated using a simple formula:

$$G(0, e) = (910 \sin(e) - 30) \text{ Wm}^{-2} \tag{4.9}$$

Their results are for Hamburg, Germany. The values 910 and 30 are very likely site dependent and need to be determined from historical data. They also reported a function for estimating the diffuse irradiance under cloud cover N octa as

$$DHI(N) = G(N)(0.3 + 0.7(N/8)^2) \tag{4.10}$$

They even suggested that given DHI and GHI, Equations 4.8 and 4.10 may be used to estimate cloud cover N in eighths.

For example, from the results from BIRD and REST2 models in Chapter 2 for Barcelona, and using Equation 4.9 (with $Zr = 32.5°$, $e = 57.5°$),

$$\text{Equation 4.9 GHI}(0) = 737 \text{ Wm}^2$$

$$\text{Bird total hemispherical radiation GHI}(0)_B = 842 \text{ Wm}^2$$

$$\text{REST total hemispherical radiation GHI}(0)_R = 878 \text{ Wm}^2$$

From Equation 4.8 and N = 4 octas,

$$G(4) = 737*(1 - 0.75(4/8)^{3.4}) = 684 \text{ Wm}^2$$

Using Equation 4.10 to estimate the diffuse irradiance under N = 4 octas of cloud cover,

Using the result from Equation 4.8:

$$DHI(4) = 684*(0.3 + 0.7(0.5^2)) = 325 \text{ Wm}^2$$

$$\text{From Bird: DHI}(4) = 399 \text{ Wm}^2$$

$$\text{From REST2: DHI}(4) = 417 \text{ Wm}^2$$

Note that this 28% range of diffuse irradiance results from the range of the clear sky models since the factor modifying GHI(0) has the same value. The Bird and REST estimates differ by 18%. For a value of DHI of 400 Wm^2, a 20% error represents about ±80 Wm^2.

As with many other authors, the classic Kasten and Czeplak article also attempted to use more information, such as cloud types (cumulus, cirrus, altostratus, cumulonimbus, etc.) to improve the accuracy of estimates. This requires detailed meteorological obser-vation, often difficult to obtain, and thus is beyond the scope of our discussion here.

4.3.2 SIMPLE CLOUD COVER MODIFIER FOR CLEAR SKY MODELS

I derived a very simple cloud cover modifier based on a relationship between global solar radiation G, solar elevation (zenith) angle e(z), and cloud cover N as reported by

TABLE 4.2
Ehnberg-Bollen Coefficient Table for Equation 4.11

N	a_0	a_1	a_3	A	L
0	−112.6	653.2	174.0	0.73	−95.0
1	−112.6	686.5	120.9	0.72	−89.2
2	−107.3	650.2	127.1	0.72	−78.2
3	−97.8	608.3	110.6	0.72	−67.4
4	−85.1	552.0	106.3	0.72	−57.1
5	−77.1	511.5	58.5	0.70	−45.7
6	−71.2	495.4	−37.9	0.70	−33.2
7	−31.8	287.5	94	0.69	−16.5
8	−13.7	154.2	64.9	0.69	−4.1

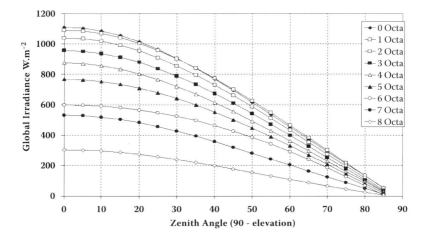

FIGURE 4.1   Ehnberg and Bollen global irradiance versus zenith angle and cloud cover N.

Ehnberg and Bollen [24]. They derived coefficients a0, a1, a3, a, and L as functions of cloud cover N in octas derived from the work of Nielsen et al. [25]. These were used to compute global radiation G from solar elevation e and cloud cover fraction, according to Equation 4.11:

$$G = [a_0(N) + a_1(N)\sin e + a_3(N)\sin^3 e - L(N)]/a(N) \qquad (4.11)$$

where the coefficients are functions of the cloud cover N in eighths, as shown in Table 4.2.

The result of computing GHI(N, Z) where z is the zenith angle (90 − e) is shown in Figure 4.1.

Computing the ratio of the global irradiances (plotted in Figure 4.1) for each cloud cover increment and 5° zenith angle interval to that at $N = 0$, $z = 0$, we obtain a relative transmittance of each cloud cover octa as a function of zenith angle z as shown in Figure 4.2.

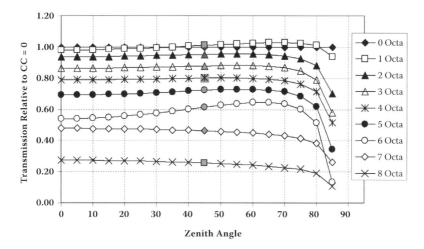

FIGURE 4.2 Cloud cover transmission curves normalized to maximum global radiation at z = 0 and N = 0 versus zenith angle. (Gray symbols at Z = 45°.)

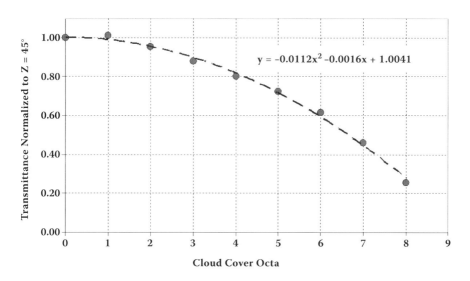

FIGURE 4.3 Cloud cover transmission versus cloud cover, in octas, for z = 45° (from gray symbols in Figure 4.2).

The transmission of each cloud amount is a relatively flat function of z, out to z = 80°. A fit of the transmission function for each cloud amount using a quadratic fit to the transmission at z = 45° (shaded symbols in Figure 4.2) versus cloud cover (CC) amount, N as shown in Figure 4.3. The relationship between N and cloud transmittance T, T = f(z, N) from Figure 4.3 can now be used to compute a transmittance cloud cover modifier (CCM) to apply to a clear sky GHI to produce a GHI under cloud cover N eighths.

Applying this transmittance function,

$$Tc = 0.0112N^2 - 0.0016N + 1.0041 \qquad (4.12)$$

For integer N = 1 to 8, values would produce GHI values that vary stepwise with N. Therefore, a scaled random number x (from a uniform distribution, $0 \le x \le 1$) is added into Equation 4.12. This introduces a little random variability into the transmittance. A scale factor that seems to work well is 0.15 x, resulting in Tc given by

$$Tc = -0.0112N^2 - 0.0016N + 1.0041 + 0.15x \qquad (4.13)$$

and GHI(N) as a function of N:

$$GHI(N) = GHI(0)(-0.0112N^2 - 0.0016N + 1.0041 + 0.15x) \qquad (4.14)$$

Myers [26,27] shows this approach works well for the Bird clear sky model, described in Chapter 3. However, for the REST2 model of Gueymard and a clear sky model by Iqbal [28], the application of the CCM results in modeled values consistently less than measured values (bias differences) of about 50 Wm^{-2} low for monthly mean daily totals of GHI.

For those models, an additional ground-to-cloud-to-ground reflection factor was developed based on some general assumptions about cloud reflectance. These assumptions are

- A reasonable typical cloud optical depth (COD) of $\tau_{diam} = 2 \sigma r_e = 25.0$
- A single scattering albedo (G) of 0.85
- An "extrapolation length" for mean free (photon) path χ of 0.58

Then, from Equation 3.138 in the work of Marshak and Davis [14], cloud reflectance Rc as function of these parameters, and cloud transmittance T_c (from Equation 4.13) is (see [14] for details):

$$Rc = T_c[(1 - G)\tau_{diam}/[2\chi] \qquad (4.15)$$

Substituting the values

$$Rc \sim (0.15*25)/(2*0.58)\, T_c = 3.23\, T_c$$

Note the value of 3.23 implies that clouds can *enhance* through reflection. This effect does in fact occur, especially in conditions of scattered clouds (see Emck and Richter [29]). An example partly cloudy day (5 May 2010 in Golden, CO) with cloud reflection enhanced GHI is shown in Figure 4.4.

Rc is cloud reflectance for radiation reflected from the ground. This is the irradiance transmitted by the clouds multiplied by the albedo of the ground ρ, generally assumed to be about 0.2. The irradiance reflected from the ground GHI(Rg) to the cloud and back to the ground GHI(Rc) becomes

$$GHI(Rc) = GHI(Rg)\, Rc = GHI(Rg)\, 3.2Tc(N/8) \qquad (4.16)$$

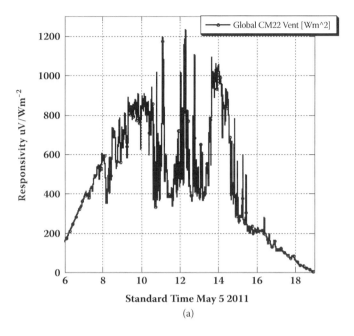

(a)

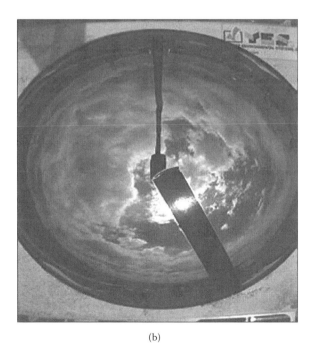

(b)

FIGURE 4.4 Example of enhanced GHI irradiance in partly cloudy conditions. All sky image at 11:00 a.m. at right (south to bottom, west to left of image). Note bright edges of cumulonimbus cloud edges near the sun.

where Tc is from Equation 4.7. The term (N/8) relates to the proportion of the sky covered by the clouds available to reflect the radiation. GHI(Rg) = GHI(N)*ρ, where GHI(N) is from Equation 4.14. Finally, we have

$$GHI = GHI(0)\ Tc + GHI(0)\ Tc\ \rho\ 3.23\ (N/8)$$

or

$$GHI = GHI(0)Tc(1 + 3.23\rho(N/8)) \tag{4.17}$$

Example: From the results from BIRD and REST2 models in Chapter 3, for Barcelona,

$$\text{Bird total hemispherical radiation GHI(0)} = 842\ \text{Wm}^2$$

$$\text{REST total hemispherical radiation GHI(0)} = 878\ \text{Wm}^2$$

Using a cloud coverage of N = 4/8, compute the GHI(4) for these results.
    The cloud transmittance from Equation 4.13 for Bird is

$$Tc = -0.0112N^2 - 0.0016N + 1.0041 = -0.0112(16) - 0.0016(4) + 1.0041 = 0.82$$

Generate a random number from 0 to 1 (e.g., 0.18); multiply by 0.1 = 0.03; add to 0.82 = 0.85 transmittance of 85%.

$$GHI(0) = 842, \text{ so } GHI(N = 4) = 842 \times 0.85 = 716\ \text{Wm}^2$$

For REST2, using the same Tc calculation,

$$GHI(N = 4) = 878*0.85 = 746\ \text{Wm}^2$$

Compute the ground-to-cloud-to-ground additional radiation (because the REST model generally underpredicts using the simple Tc):

Assume ρ = 0.2, so the upward reflected ground radiation is 0.2 × 746 = 149 Wm².
The cloud reflectance for N = 4 is 3.23*0.85*(4/8) = 1.37

Multiplying the ground-reflected radiation of 149.0 Wm² by the cloud reflectance 1.37 = 204 additional watts, and the resultant REST GHI(4) irradiance = 764 + 204 = 968 Wm².
    The 250-W difference between these two results is mainly due to using the cloud-to-ground reflectance correction for the REST2 model. This correction was developed to account for a bias in the long-term time series of REST2 calculations when using the cloud cover transmittance modifier alone.
    The simple cloud cover modifier does not account for the spatial or vertical distribution of the clouds. It does not even accurately reflect whether the direct beam is blocked by clouds.

The differences between various clear sky models may be large for any single-point calculation.

Over longer periods of time such as months, random differences (induced by the cloud cover modifier) between model results should approximately cancel out. Then, monthly mean daily totals or monthly hourly average or monthly daily average modeled irradiances should approach differences represented by the systematic differences in the clear sky models. This is an important point to keep in mind with every solar radiation model. Mean bias errors of about –7% (underestimation of the irradiance) and random (root mean square) errors on the order of 20% to 30% in the modeled hourly data have been demonstrated for the general cloud cover modifier described here [26,27].

The diurnal pattern of cloudiness also drives the diurnal pattern of the solar radiation profile. Figure 4.5 shows diurnal profiles of solar GHI radiation for Albany, New

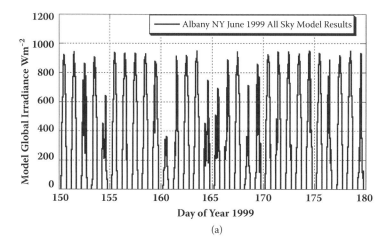

(a)

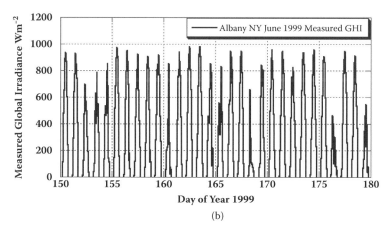

(b)

FIGURE 4.5 Time series of June 1999, Albany, New York, GHI solar radiation modeled (a) from Bird clear sky model with CCM and measured GHI data (b) for the same period.

York, for June 1999 using cloud cover observations and the CCM applied to the Bird clear sky model compared with measured GHI data for the same period. The important observation is that the profiles derived from the cloud cover observations result in a realistic representation of the natural variability in the solar radiation.

4.4   ALL SKY SOLAR RADIATION FROM WEATHER SATELLITES

The idea of deriving solar radiation data from weather satellite images emerged soon after weather satellite images became routinely available [30]. As of 2012, two popular models have been developed and applied worldwide. These models are Heliostat-2, developed in the European Community, based on the METEOSAT (Meteorological Satellite) series of satellites [31]. Somewhat later, Perez et al. [32] developed the so-called SUNY (State University of New York) satellite model based on the U.S. GOES (Geosynchronous Orbiting Environment Satellite) series of satellite images.

These models involve complex manipulation of the raw satellite images. Image pixel brightness with respect to "clear sky" pixel brightness "calibration" is required. Next, the effects of clouds, in terms of a "cloud index" or some other cloud transmittance function, are computed. Using assumptions or databases of atmospheric transmittance (aerosol optical depth in particular) in combination with clear sky models and solar geometry (for each pixel), a global hemispherical irradiance on the horizontal is computed. Next, so-called decomposition models described in the next chapter are used to estimate direct beam and diffuse irradiance. These parameters are all computed over a defined grid of specified resolution, such as 10×10 km. The temporal resolution of the satellite-based data sets may require interpolation or correction, depending on the timing of the satellite images and the desired data time step size.

The accuracy of satellite-derived data is generally assessed by comparisons with ground measurement stations. A summary of several evaluations of these models was provided provided by Gueymard in Chapter 18, Section 5, of Badescu's work [33]. The results for global hemispherical irradiance estimates are summarized in Table 4.3.

Table 4.3 shows that hourly bias or systematic error for GHI can be made rather small, less than 5%, and that there is considerable random scatter in the correlation of the satellite results with ground data. As direct irradiances are generally derived

TABLE 4.3
Examples of Satellite Model Accuracy Assessments

Author [Reference]	Date	Component	Hourly RMSE%	Hourly MBE%
Zelenka et al. [34]	1999	Global	23.0	N/A
Martins et al. [35]	2003	Global	25.6	2.7
Rigollier et al. [31]	2004	Global	19.9	−2.7
Shillings et al. [36]	2004	Beam	36.1	4.3
Zarzalejo et al. [37]	2005	Global	18.3	0.27

from GHI, both types of errors are greater for this component. Chapter 6 describes this approach in more detail.

4.5 THE FUTURE: FORECASTING SOLAR RADIATION

At the time of this writing (2012), forecasting of solar radiation for a few hours to a few days in advance was just beginning to be investigated. Some statistical approaches were available based on correlation studies of solar radiation and "cloud vector motion" derived from sequential (historical; previous few hours) cloud images from space [38] or numerical weather forecast (NWF) models that predict cloud and weather patterns. The last-mentioned models are used to produce the forecasts used for nightly television weather forecasting. These include the U.S. National Digital Forecast Database (NDFD) comprised of elements of models such as the Weather Research Forecast (WRF) model and European Center for Medium Range Weather Forecast (ECMWF) model [39]. These models are more complex and require special knowledge to operate in a consistent manner. The state of the art of these models is such that day-ahead forecasts have mean bias errors on the order of 10% to 15% and root mean square (RMSE) errors on the order of 40% with respect to measured data. So-called ensemble modeling, consisting of averages of multiple results can sometimes cut these errors in half. However, it is clear there is much room for improvement in these models.

4.6 SUMMARY

Modeling solar radiation under all sky conditions is considerably more difficult and uncertain than modeling clear sky solar radiation. The physics of cloud transmittance and the extremely wide variety of spatial distributions of clouds are the main contributors to these difficulties. Resulting uncertainties in model data are presently about 3% to 5% in systematic or bias errors and about 20% to 30% in random errors, or RMSEs, depending on the component of interest and model used. Either very detailed ground-based observations or classification of cloud properties or perhaps more complex, larger-scale "ensemble" or statistical modeling techniques are required to improve the state of the art in this regime.

In the meantime, simple correlations of measured solar data and cloud cover estimates are the most straightforward engineering tools for estimating global radiation in arbitrary cloud cover conditions. Modeling of solar radiation based on satellite images of cloud cover in conjunction with estimated or derived atmospheric properties requires technical resources usually beyond the resources of the individual or even a small team of engineers. The same techniques used for weather satellite-based observations are just beginning to be applied to numerical weather forecasting models (where cloud patterns are predicted a few hours to a few days in advance) for forecasting solar radiation. For a digital atlas of worldwide cloud types, their frequency of occurrence, optical properties, and publications, visit http://www.atmos.washington.edu/~ignatius/CloudMap (accessed 21 July 2012). It is possible that a combination of cloud physics and empirical model results will be used for forecasting solar radiation in the future but is beyond our scope here.

REFERENCES

1. Hargreaves, G.H., and Z.A. Samani. (1982). Estimating potential evapotranspiration. *Journal of Irrigation and Drain Engineering, ASCE*, Vol. 108 (IR3), pp. 223–230.
2. Hargreaves, G.H., and Z.A. Samani. (1985). Reference crop evapotranspiration from temperature. *Transactions of ASAE,* Vol. 1, No. 2, pp. 96–99. http://cagesun.nmsu.edu/~zsamani/papers/Hargreaves_Samani_85.pdf.
3. Knapp, C.L., T.L. Stoffel, and S.D. Whitaker. (1980). *Insolation Solar Radiation Manual*. Solar Energy Research Institute, Golden, CO. 281 pp.
4. Allen, R.G. (1995). *Evaluation of Procedures for Estimating Mean Monthly Solar Radiation from Air Temperature*. Report submitted to the United Nations Food and Agricultural Organization (FAO), Rome Italy. http://cagesun.nmsu.edu/~zsamani/research_material/files/Hargreaves-samani.pdf.
5. Angstrom, A. (1924). Solar and terrestrial radiation. *Quarterly Journal of the Royal Meteorological Society*, Vol. 50, p. 121.
6. Prescott, J. A. (1940). Evaporation from a water surface in relation to solar radiation. *Transactions of the Royal Society of South Australia*, Vol. 64, pp. 114–118.
7. Angstrom, A. (1956). On the computation of global radiation from records of sunshine. *Arkiv for Geofysik,* Vol. 2, No. 22, p. 471.
8. Suehrcke, H. (2000). On the relationship between duration of sunshine and solar radiation on the earth's surface: Angstrom's equation. *Solar Energy*, Vol. 68, No. 5, pp. 417–425.
9. Gueymard C., P. Jindra, and V. Estrada-Cajigal. (1995). A critical look at recent interpretations of the Angstrom approach and its future in global solar radiation prediction. *Solar Energy*, Vol. 54, p. 357.
10. World Meteorological Organization. (2008). *WMO Guide to Meteorological Instruments and Methods of Observation,* WMO No. 8, 7th ed. World Meteorological Organization, Geneva, Switzerland. http://www.wmo.int/pages/prog/gcos/documents/gruanmanuals/CIMO/CIMO_Guide-7th_Edition-2008.pdf.
11. Glover, J., and J. McCulloch. (1958). The empirical relation between solar radiation and hours of bright sunshine. *Quarterly Journal of the Royal Meteorological Society*, Vol. 84, p. 172.
12. Almorox, J., M. Benitob, and C. Hontoria. (2005). Estimation of monthly Angstrom-Prescott equation coefficients from measured daily data in Toledo, Spain. *Renewable Energy,* Vol. 30, pp. 931–936.
13. Tymvios, F.S., C.P. Jacovides, S.C. Michaelides, and C. Scouteli. (2005) Comparative study of Angstrom's and artificial neural networks' methodologies in estimating global solar radiation. *Solar Energy*, Vol. 78, No. 6, pp. 752–762.
14. Marshak, A., and A.B. Davis (eds.). (2005) *3D Radiative Transfer in Cloudy Atmospheres*. Springer, Berlin.
15. Hahn, C.J., S.G. Warren, J. London, R.M. Chervin, and R. Jenne. (1982). Atlas of Simultaneous Occurrence of Different Cloud Types over the Ocean. NCAR Technical Note NCAR/TN-201+STR. National Center for Atmospheric Research, Boulder, CO.
16. Garcia, O. (1985). Atlas of Highly Reflective Clouds for the Global Tropics, 1971–1983. U.S. Department of Commerce, Boulder, CO; National Oceanic and Atmospheric Administration, Environmental Research Laboratories, Washington, DC.
17. Smith, G.L., K.J. Priestley, N.G. Loeb, B.A. Wielicki, T.P. Charlock, P. Minnis, D.R. Doelling, and D.A. Rutan. (2011). Clouds and Earth Radiant Energy System (CERES), a review: Past, present and future. *Advances in Space Research*, Vol. 48, pp. 254–263.
18. Chambers, L., B. Wielicki, and K. Evans. (1997). Accuracy of the independent pixel approximation for satellite estimates of oceanic boundary layer cloud optical depth. *Journal of Geophysical Research-Atmospheres*, Vol. 102(D2), pp. 1779–1794.

19. Rossow, W.B., and R.A. Schiffer. (1991). ISCCP cloud data products. *Bulletin of the American Meteorological Society*, Vol. 71, pp. 2–20.
20. Kasten, F., and G. Czeplak. (1980). Solar and terrestrial radiation dependent on the amount and type of cloud. *Solar Energy*, Vol. 24, No. 2, pp. 177–189.
21. Gul, M.S., T. Muneer, and H.D. Kambezidis. (1998). Models for obtaining solar radiation data from other meteorological data. *Solar Energy*, Vol. 64, No. 1–3, pp. 98–108.
22. Chen, R., E. Kang, X. Ji, J. Yang, and J. Wang. (2007). An hourly solar radiation model under actual weather and terrain conditions: A case study in Heihe river basin. *Energy*, Vol. 32, pp. 1148–1157.
23. Muneer, T. (2004). *Solar Radiation and Daylight Models*, 2nd ed. Elsevier, Butterworth, Heineman, Oxford, UK.
24. Ehnberg, J.S.G., and M.H.J. Bollen. (2005). Simulation of global solar radiation based on cloud observations. *Solar Energy*, Vol. 78, pp. 157–162.
25. Nielsen, L. B., I.P. Prahm, R. Berkowicz, and K. Conradsen. (1981). Net incoming radiation estimated from hourly global radiation and/or cloud observations. *Journal of Climatology*, Vol. 1, No. 3, pp. 255–272.
26. Myers, D. (2006). Cloudy sky version of Bird's broadband hourly clear sky model. In Solar 2006, *Proceedings of American Solar Energy Society and American Society of Mechanical Engineers Joint Conference*, 9–13 July 2006, Denver, CO. American Solar Energy Society, Boulder, CO.
27. Myers, D. (2007). General cloud cover modifier for clear sky solar radiation models. *Proceedings of the SPIE, Optical Modeling and Measurements for Solar Energy Systems*, Vol. 6652, San Diego CA, 26–28 August 2007. Society of Photo-Optical Instrumentation Engineers, Bellingham, WA.
28. Iqbal M. (1983). *An Introduction to Solar Radiation*. Academic Press, Toronto.
29. Emck, P., and M. Richter. (2008). The upper threshold of global shortwave irradiance in the troposphere derived from field measurements in tropical mountains. *Journal of Applied Meteorology and Climatology*, Vol. 27, pp. 2828–2845.
30. Dubayah, R., and S. Loechel. (1997). Modeling topographic solar radiation using GOES data. *Journal of Applied Meteorology*, Vol. 36, pp. 141–154.
31. Rigollier, C., M. Lefèvre, and L. Wald. (2004). The method Heliostat-2 for deriving shortwave solar radiation from satellite images. *Solar Energy*, Vol. 77, No. 2, pp. 159–169.
32. Perez, R., P. Ineichen, K. Moore, M. Kmiecik, C. Chain, R. George, and F. Vignola. (2002). A new operational model for satellite derived irradiances: Description and validation. *Solar Energy*, Vol. 73, No. 5, pp. 307–317.
33. Badescu, C. (ed.). (2008). *Modeling Solar Radiation at the Earth Surface*. Springer, Berlin.
34. Zelenka, A., R. Perez, R. Seals, and D. Renne. (1999). Effective accuracy of satellite derived hourly irradiances. *Theoretical and Applied Climatology*, Vol. 62, pp. 199–207.
35. Martins, F.R., B.E. Pereira, S.L. Abreu, H.G. Beyer, S. Colle, R. Perez, and D. Heinemann. (2003). Cross validation of satellite radiation transfer models during SWERA project in Brazil. *Proceedings ISES 2003*, Göteborg, 14–19 June. International Solar Energy Society, Freiburg, Germany.
36. Shillings, C., R. Meyer, and H. Mannstein. (2004). Validation of a method for deriving high resolution direct normal irradiance from satellite data. *Solar Energy*, Vol. 76, pp. 485–497.
37. Zarzalejo, L.F., L. Ramirez, and J. Polo. (2005). Artificial intelligence techniques applied to hourly global irradiance estimation from satellite derived cloud index. *Energy*, Vol. 30, pp. 1687–1697.
38. Hamill, T. M., and T. Nehrkorn. (1993). A short term cloud forecast scheme using cross correlations. *Weather and Forecasting*, Vol. 8, No. 4, pp. 401–411.
39. Perez, R., M. Beauharnois, K. Hemker, Jr., S. Kivalov, E. Lorenz, S. Pelland, J. Schlemmer, and G. Van Knowe. (2011). Evaluation of numerical weather prediction solar irradiance forecasts in the US. *Proceedings Solar 2011*, ASES National Conference, Raleigh, NC. American Solar Energy Society, Boulder, CO.

5 Modeling Missing Components

Without analysis, no synthesis.

—Friedrich Engels, 1878

5.1 INTRODUCTION

Typically, not all three solar radiation components are available as measured data. Global horizontal irradiance (GHI) is the most common measurement because it uses the simplest measurement equipment: a pyranometer on a horizontal surface. Because of the complexity of equipment for measuring direct normal irradiance (DNI) radiation, this is the most common missing measurement. Measuring diffuse horizontal irradiance (DHI) irradiance with shading bands is not complex or very expensive, except that the data must be corrected for the obscuration of the sky by the shading band. DHI measurements using tracking shading devices (disks or balls) can be as expensive as DNI measurements because of the solar-tracking requirements.

Chapter 3 described the modeling of all three radiation components for clear sky situations, and Chapter 4 did so for all sky conditions. The accuracy of the estimates based on just the model approaches described may be improved if at least one measured component, usually GHI, is available. Many authors have derived so-called decomposition models, which use empirical relationships between GHI and the other two components to estimate the missing components. A further application of these decomposition models is to use a modeled GHI or DNI estimate to derive the complementary components and compare with those derived from the more complete model algorithms. This provides an estimate of the relative uncertainty or accuracy of the two approaches to deriving all three model components.

5.2 ESTIMATING DIFFUSE FROM GLOBAL HORIZONTAL IRRADIANCE

The diffuse sky radiation on a horizontal surface DHI is a function of cloud cover, the scattering properties of the atmosphere, and the path length (air mass M, dependent on solar elevation or zenith angle). Since detailed information on the state of the atmosphere and cloud type, altitude, and spatial distribution is not usually available, correlations of DHI with GHI have been studied with the goal of estimating DHI. The parameterizations discussed here were based on estimating hourly average DHI and used the ratio of hourly DHI to GHI as the dependent variable. As can be

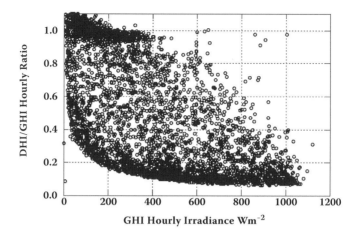

FIGURE 5.1   Plot of DHI/GHI ratio as function of GHI (NREL SRRL 2005).

imagined, and as Figure 5.1 demonstrates, for any given hourly GHI value, a wide range of hourly DHI and DHI/GHI ratio values can occur.

Given this situation, researchers separate various regimes of DHI/GHI relationships based on additional variables. The most popular additional parameter, easily derived from the GHI data, is the global hemispherical clearness index Kt, described in Chapter 1, Section 1.5, Equation 1.24.

5.2.1   ORGILL AND HOLLANDS CORRELATION

The first correlation study examining the relationship of DHI to GHI as a function of clearness index Kt was that of Orgill and Hollands [1]. Their study was based on 4 years of Canadian measured GHI and DHI (using a pyranometer under a shadow band) solar radiation data from Toronto, Canada.

Recall that Kt = GHI/(Io cos(z)). The authors partitioned Kt into three ranges representing clear, partly cloudy, and very cloudy regimes. For each Kt range, the ratio of DHI/GHI was fit with a linear function, DHI/GHI = a + b Kt. The resulting fits were

$$DHI/GHI = 0.177 \qquad\qquad Kt > 0.75 \qquad\qquad (5.1)$$

$$= 1.577 - 1.84\ Kt \qquad 0.35 \leq Kt \leq 0.75 \qquad (5.2)$$

$$= 1.0 - 0.249\ Kt \qquad\quad 0 \leq Kt < 0.35. \qquad (5.3)$$

5.2.2   ERBS CORRELATION

Erbs et al. [2] performed higher-order polynomial fits of DHI/GHI ratios using data from the United States (i.e., for a range of lower latitudes than Orgill and Hollands). They used measured DNI and GHI data to compute DHI [= GHI − DNI Cos (z)] for stations with 1 to 4 years of data. They also partitioned the Kt parameter into three regimes; however, the range for each regime is different from the Orgill and Hollands ranges. The resulting polynomial fit equations were

$$DHI/GHI = 0.165 \qquad\qquad\qquad\qquad\qquad Kt > 0.80 \qquad (5.4)$$

$$= 0.951 - 0.160Kt + 4.388Kt^2 - 16.64Kt^3 + 12.34Kt^4 \quad 0.22 \le Kt \le 0.80 \quad (5.5)$$

$$= 1.0 - 0.09Kt \qquad\qquad\qquad\qquad\qquad 0 \le Kt < 0.22 \quad (5.6)$$

5.2.3   BOES DNI CORRELATION

Since the GHI is composed of a combination of DNI and DHI, it is reasonable to assume that DHI can be obtained by using a correlation of DNI with GHI and obtain DHI by subtraction. This is the approach of Boes et al. [3]. Their study used 1 year of data from three U.S. stations with measured DNI and GHI data. Their correlation produces DNI (in Wm^2) as a function of Kt and solar elevation angle:

$$DNI = 400 \; Wm^2 \qquad z \ge 80° \qquad Kt > 0.5 \qquad (5.7)$$

$$= -520 + 1.8Kt \qquad 0.3 \le Kt \le 0.85 \qquad (5.8)$$

$$= 1000 \; Wm^2 \qquad\qquad Kt > 0.85 \qquad (5.9)$$

The original work reported monthly values of clear sky DNI ($Kt > 0.85$) ranging from 950 to 1050 Wm^2. Here, we simply use the value of 1000 Wm^2 to simplify the model.

DHI is then computed as GHI – DNI cos(z).

Given the simplicity of these models, the reader can compute and compare results for several different combinations of GHI, DNI, and DHI. However, for the data shown in Figure 5.1, we show in Figure 5.2 a histogram of the percentage differences between the measured DHI/GHI ratio and that computed using the Orgill and Hollands algorithm.

The average DHI magnitude for the 400 data points that exceeded 90% difference between the measured and modeled ratio of DHI/GHI was 36 Wm^2, representing a low DHI/GHI ratio, whether measured or modeled, and thus the ratios of small numbers. These data represent about 10% (400 h of 4280 h of daylight) of the total data set. The histogram shows that the Orgill algorithm was accurate to about ±20% for estimating DHI from GHI most of the time.

5.3   ESTIMATING DIRECT FROM GLOBAL NORMAL IRRADIANCE

Just as correlations have been developed for estimating diffuse irradiance, similar research has addressed the estimation of DNI from measured GHI data. The Boes correlation of Section 5.2.3 introduced this concept and is a very simple correlation. Analogous to Figure 5.1, Figure 5.3 plots the ratio of hourly DNI/GHI irradiance versus GHI irradiance for the same station year.

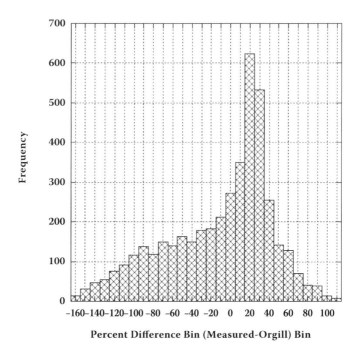

FIGURE 5.2   Histogram of measured and Orgill and Hollands estimated DHI/GHI ratio for the hourly data in Figure 5.1.

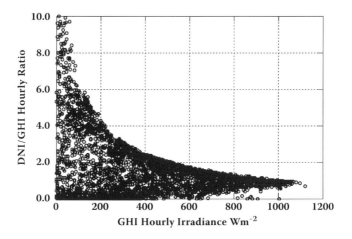

FIGURE 5.3   Ratio of hourly DNI/GHI versus GHI irradiance for same site as Figure 5.1.

5.3.1 THE MAXWELL DIRECT INSOLATION SIMULATION CODE MODEL

Just as other authors have partitioned the DHI/GHI into different functions for different Kt regimes and Boes et al. [3] did for the DNI/GHI ratio, Maxwell [4] used the same approach for correlating DNI/GHI in his direct insolation simulation code (DISC) model. Rather than use the DNI/GHI ratio for the dependent variable, or output of the model, he used Kn, the direct beam clearness index. Maxwell developed the model using 1 year of measured data for Atlanta, Georgia, in the United States. Validation of the model was performed by applying the model to three U.S. sites with diverse climates: Brownsville, Texas; Albuquerque, New Mexico; and Bismarck, North Dakota. Figure 5.4 shows the relationship between Kn and Kt for the same site as Figures 5.1 and 5.2. Note that this relationship will vary from site to site depending on the radiation "climate" for each site.

As usual, for any given value of Kt (GHI), several different values of Kn (DNI) and Kd (DHI) can combine to produce the measured Kt or GHI.

As described by Maxwell [4], parameterizations of Kn with respect to bins of width 0.05 and air mass were studied and combined to result in a general formulation of Kn versus Kt. The model equation produces the change or deviation ΔKn from clear sky Kn, denoted by Knc. Modeled DNI is computed as

$$DNI = Io\ Kn \qquad (5.10)$$

where

$$Kn = Knc - \Delta Kn \qquad (5.11)$$

and

$$\Delta Kn = a + be^{(C*AM)} \qquad (5.12)$$

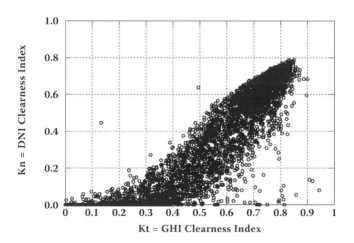

FIGURE 5.4 Data for Figure 5.2 transformed into "K space" or clearness indices.

Clear sky limit Knc is a polynomial in air mass AM:

$$Knc = 0.866 - 0.122AM + 0.0121AM^2 - 0.000653AM^3 + 0.000014AM^4 \quad (5.13)$$

The Kt space is partitioned into two parts, Kt ≤ 0.6 and Kt > 0.6:

For Kt ≤ 0.60,

$$a = 0.512 - 1.56Kt + 2.286Kt^2 - 2.222Kt^3 \quad (5.14)$$

$$b = 0.370 + 0.962Kt^3 \quad (5.15)$$

$$c = -0.280 + 0.932Kt - 2.048Kt^2 \quad (5.16)$$

For Kt > 0.60,

$$a = -5.743 + 21.77Kt - 27.49Kt^2 + 11.56Kt^3 \quad (5.17)$$

$$b = 41.4 - 118.5Kt + 66.05Kt^2 + 31.90Kt^3 \quad (5.18)$$

$$c = -47.01 + 184.2Kt - 222.0Kt^2 + 73.81Kt^3 \quad (5.19)$$

The model was shown to have bias errors of about ±50 Wm² and random errors of about ±150 Wm² depending on the site.

There is one caution about using the DISC universally, and that is that the range of zenith angles, or air masses for the model development and validation, are only for continental sites with latitude ranging from 28 N to 45 N. While many evaluations of the model have been accomplished, it is not entirely clear that the model will perform well at any site outside this latitude range.

5.3.2 THE PEREZ **DIRINT** AND **DIRINDEX** MODELS

5.3.2.1 **DIRINT** Model

As mentioned in the previous section, Perez et al. [5] and others [6] have evaluated the DISC model. Most evaluations resulted in the range of bias and random errors mentioned at the end of that section. In an effort to improve the performance of DISC, Perez worked with the American Society of Heating Refrigerating, and Air Conditioning Engineers (ASHRAE) to develop versions of DISC, called DIRINT and DIRINDEX. The DIRINT model uses a variation of the Kt parameter called Kt′ (Kt prime) developed earlier by Perez et al. [7] that is independent of zenith angle:

$$Kt' = Kt/(1.031 \cdot e^{[-1.4/(0.9+9.4/AM)]} + 0.1) \quad (5.20)$$

DIRINT also introduces a "stability index" (ΔKt′) parameter that changes with time:

$$\Delta Kt' = 0.5 \cdot (\mid Kt'_i - Kt'_{i+1} \mid + \mid Kt'_i - Kt'_{i-1} \mid) \tag{5.21}$$

where i, i + 1, and i − 1 refer to the current, next, and previous hourly record, respectively. Utilizing the three parameters zenith angle, Kt′, and ΔKt′, DIRINT was characterized as a three-dimensional DIRINT model (called "3-D ΔKt" in their article).

Adding an additional input variable (hourly dew point temperature Td), they examined the performance of a "4-D" model as well. Multidimensional binning of solar irradiance data with respect to Kt′, ΔKt′, water vapor (W), derived from Td if available, and zenith angle (Z) resulted in a $6 \times 6 \times 5 \times 7$ four-dimensional (4-D) lookup table. The table was based on six Kt′ and Z bins each, five bins for W, and seven bins for ΔKt′ described in Table 1 of Perez et al. [5]. In this version, In = I_{disc} · X (Kt′, Z, W, ΔKt′), where I_{disc} was computed directly from the DISC algorithm and was modified by looking up the correction coefficient X.

The second (4-D) version of the DIRINT model computed the direct irradiance using

$$In = I_0 \cdot Kb' \cdot e^{\,[-1.4/(0.9+9.4/AM)]}/0.87291 \tag{5.22}$$

where Kb′ = 0 if Kt′ < 0.2; otherwise, Kb′ = a Kt′ + b, where a and b are functions of Kt′, Z, W, and ΔKt′ in an $8 \times 5 \times 4 \times 6$ lookup table with bins similar to those described previously. The appropriate version (3-D if Td not available, 4-D if Td available) of the model computes DNI.

The DIRINT model performs only slightly better than the DISC model with the following input variables: hourly GHI data; hourly solar zenith angle; daily value of solar constant; altitude of location; and hourly Td (if available); as well as the lookup matrices for the correction coefficients appropriate for the 3-D or 4-D model version.

5.3.2.2 DIRINDEX Model

The DIRINDEX model modifies the DIRINT model result with information from the Bird model or another clear sky model. The DIRINDEX uses the hourly clear sky GHI and DNI data from the Bird clear sky model described in Chapter 3, Section 3.4.1.

The DNI for the DIRINDEX model was calculated by the following equation:

$$DNI_{DIRINDEX} = [(DNIclear)(DIRINT(GHI))]/DIRINT(GHIclear) \tag{5.23}$$

where
   DIRINT(GHI) = DNI calculated with DIRINT model with measured GHI as input data (Wm^2)
   DIRINT(GHIclear) = DNI calculated with the DIRINT model with calculated GHI from a clear sky model if AOD and water vapor known with confidence
   DNIclear = DNI calculated with a clear sky model (Wm^2)

Table 5.1 shows a comparison of the performance of DISC, DIRINT and DIRINDEX in terms of mean bias error (MBE) and root mean square error (RMSE) for three southwestern U.S. sites in 2005 [8]. The sites are the National Renewable Energy Laboratory Solar Radiation Research Laboratory (SRRL) in Golden,

TABLE 5.1

Monthly Mean Bias (MBE) and Root Mean Square (±RMSE) Percentage Errors of DNI for Three Locations and Three GHI-to-DNI Conversion Models Based on the Maxwell DISC Model Using Hourly Measured Data from 2005

Location Model	Error (%)	Jan	Feb	Mar	Apr	May	Jun	Jul	Aug	Sep	Oct	Nov	Dec
CPRL	Monthly Mean Daily Wm²	2572	5614	6529	6626	5355	7820	7946	5616	6896	5777	6728	5542
DISC	MBE	-0.3	0.9	1.6	0.8	0.5	-0.1	-0.1	0.1	0.8	0.1	0.7	1.1
DIRINT	MBE	-0.2	0.2	1.0	0.5	0.3	0.0	-0.2	-0.5	0.0	-0.2	-0.2	0.1
DIRINDEX	MBE	0.3	0.4	0.7	0.2	0.5	-0.2	-0.2	-0.2	0.2	-0.1	0.0	0.7
DISC	RMSE	8.0	3.5	3.0	2.0	2.2	1.6	1.4	2.0	2.1	2.4	2.0	3.1
DIRINT	RMSE	3.5	1.7	1.6	1.1	1.5	1.4	1.1	1.7	1.2	1.2	1.3	2.3
DIRINDEX	RMSE	3.6	1.6	1.9	1.1	1.8	1.5	1.1	1.8	1.4	1.2	1.2	2.3
PSEL	Monthly Mean Daily Wm²	5758	4444	6272	7664	8293	9255	8264	6542	6543	6359	6886	6244
DISC	MBE	0.3	1.1	0.4	-0.1	-0.6	-0.8	-0.8	-1.3	-1.7	-0.3	-1.1	0.2
DIRINT	MBE	-0.2	0.7	0.0	-0.2	-0.3	-0.5	-0.6	-1.1	-0.9	-0.6	-0.9	-0.1
DIRINDEX	MBE	0.2	0.9	0.0	-0.2	-0.3	-0.6	-0.6	-0.9	-0.8	-0.5	-0.2	0.7
DISC	RMSE	3.0	3.5	2.3	1.6	1.4	1.4	1.7	1.8	2.2	1.8	2.5	2.2
DIRINT	RMSE	1.6	2.3	1.4	0.9	0.7	0.7	1.0	1.3	1.0	1.2	1.2	1.5
DIRINDEX	RMSE	1.5	2.4	1.4	0.9	0.8	0.7	1.0	1.3	1.1	1.2	1.2	1.7
SRRL	Monthly Mean Daily Wm²	6190	6435	6443	6049	6482	6460	7117	5755	5934	5996	5418	4973
DISC	MBE	0.5	0.9	1.3	0.6	0.0	-0.9	-0.7	-0.6	-0.8	-0.1	-0.1	0.8
DIRINT	MBE	0.0	0.4	0.7	0.2	0.1	-0.4	-0.5	-0.5	-0.4	-0.3	-0.4	0.2
DIRINDEX	MBE	0.5	0.7	0.4	0.0	0.2	-0.5	-0.5	-0.6	-0.7	-0.5	0.1	0.8
DISC	RMSE	2.9	2.9	2.5	2.1	2.4	2.2	1.9	1.9	2.4	2.1	3.1	3.8
DIRINT	RMSE	1.6	1.7	1.5	1.2	1.5	1.0	1.1	1.4	1.1	1.3	1.8	2.2
DIRINDEX	RMSE	1.8	2.0	1.6	1.2	1.7	1.0	1.2	1.5	1.2	1.4	2.0	2.4

Shaded cells indicate absolute MBE > 50 Wm² or RMSE > 150 Wm².

Colorado; the Department of Agriculture Conservation and Production Research Laboratory (CPRL), Bushland, Texas (just west of Austin, TX); and the Photovoltaic Systems Evaluation Laboratory (PSEL) at the Sandia National Laboratory at Albuquerque, New Mexico.

GHI measurements used for model input were made with a Kipp and Zonen CM 21 ventilated pyranometer (uncertainty ±4%) at SRRL, LiCor Li-200 silicon photodiode pyranometer (uncertainty ±5%) at CPRL, and Eppley precision spectral pyranometer (uncertainty ±4%) at PSEL. DNI measurements were made with a Kipp and Zonen CH1 pyrheliometer at SRRL and Eppley normal incidence pyrheliometers at CPRL and PSEL; all had an uncertainty of ±2.0%.

The model mean bias and root mean square percentage errors for monthly mean daily total DNI are reported in the table. The monthly mean values of DNI for each month for the three sites are shown to illustrate the magnitudes involved.

For all three sites and three models, the percentage MBE was within ±1.7%. The RMSEs averaged ±1.8% but ranged from ±1% to ±8% depending on the site and the month. Each site had various periods when the absolute bias errors exceeded 50 Wm^2 and the RMSE errors exceeded 150 Wm^2, represented by the shaded cells in the table. Thus, all three models had about the same performance for these sites with only minor differences.

5.4   SUMMARY

The best-possible means of ensuring the quality and consistency of solar radiation data is to have all three solar components available. This allows one to use the component balance equation to examine the consistency of each component. When only one of the three components of solar radiation is missing, it may be computed from the other two using the component balance equation. The real challenge is when only one component is measured, and we wish to estimate or derive one or both of the other two components. Since global hemispherical (total horizontal) radiation is by far the more prevalent solar data, many efforts have been devoted to converting GHI to DNI or DHI. The models described ranged from very simplistic to intricate. The influence of the variable atmospheric filter and clouds significantly expands the parameter space containing the relationship between the three components. This parameter space can be transformed from irradiance values (or the ratio of irradiances, such as DHI/GHI) to "clearness index" space, which may simplify some of the relationships, but the scatter in the data will always be there.

Thus, modeling any missing component from any one measured solar component will always be fraught with large uncertainties. The conversion of GHI to DNI using Maxwell's DISC model has not been significantly improved since 1987. The DIRINT or DIRINDEX variations of Perez have only slightly lower bias differences from measured data, smaller than the measurement uncertainty available to verify the improvement in accuracy; and the scatter or random errors in all three models are about the same. However, the DISC model was developed for only a limited geographical range of stations and thus a limited range of zenith angles. More research needs to be done to improve and expand this DISC model limitation.

REFERENCES

1. Orgill, J.F., and K.G. Hollands. (1977). Correlation equation for hourly diffuse radiation on a horizontal surface. *Solar Energy* 19, Vol. 4, pp. 357–359.
2. Erbs, D.G., S.A. Klein, and J.A. Duffie. (1982). Estimation of the diffuse radiation fraction hourly, daily, and monthly-average global radiation. *Solar Energy,* Vol. 28, No. 4, pp. 293–304.
3. Boes, E.C., H.E. Anderson, I.J. Hall, R.R. Prairie, and R.T. Stromberg. (1977). *Availability of Direct, Total and Diffuse Solar Radiation to Fixed and Tracking Collectors in the USA.* Sandia Report SAND77-0885. Solar Energy Research Institute, Golden, CO.
4. Maxwell, E.L. (1987). *A Quasi-Physical Model for Converting Global Horizontal to Direct Normal Insolation.* SERI Report SERI/TR-215-3087. http://www.nrel.gov/docs/legosti/old/3087.pdf. Accessed 21 July 2012.
5. Perez, R., P. Ineichen, E. Maxwell, R. Seals, and A. Zelenka. (1992). Dynamic global-to-direct irradiance conversion models. *ASHRAE Transactions Research,* 3578(RP-644), pp. 354–369.
6. Ineichen, P. (2008). Comparison and validation of three global-to-beam irradiance models against ground measurements. *Solar Energy,* 82, pp. 501–512.
7. Perez, R., P. Ineichen, R. Seals, and A. Zelenka. (1990). Making full use of the clearness index for parameterizing hourly insolation conditions. *Solar Energy,* Vol. 45, No. 2, pp. 111–114.
8. Vick, B.D., D.R. Myers, and W.E. Boyson. (2012). Using direct normal irradiance models and utility electrical loading to assess benefit of a concentrating solar power plant. *Solar Energy,* Vol. 86, No. 12, pp. 3519–3530.

6 Applications: Modeling Solar Radiation Available to Collectors

In time, manufacturing will to a great extent follow the sun.

—C. G. Abbot, 1928

6.1 SOLAR COLLECTOR GEOMETRIES

So far, modeling solar radiation available on horizontal surfaces only has been described. As we have seen, this resource is a combination of the direct normal beam radiation (direct normal irradiance, DNI) and its incident angle on a horizontal surface and sky diffuse radiation available from the sky dome. We have mentioned several times that ground-reflected radiation is available to surfaces tilted below the horizontal plane. Surfaces tilted at fixed angles away from horizontal and in various azimuth directions away from due south (or due north in the Southern Hemisphere) can be used to increase or optimize solar energy harvesting for many applications. Flat plate and concentrating collectors that track the sun or change tilt or azimuth of the collecting surface or aperture throughout the day can also increase energy harvesting.

Certain geometries have become more or less standard as energy-harvesting configurations based on modeling studies, "rules of thumb," or experience. Table 6.1 describes some of the most common configurations, the rationale for using them, and approximate enhancement factors with respect to energy on a horizontal surface.

Since it is relatively expensive to collect measured solar radiation in various combinations of these collector configurations, many models have been developed to convert the more typical measured (or modeled) solar radiation components (global horizontal irradiance [GHI], DNI, diffuse horizontal irradiance [DHI]) to the radiation available for these (or any other arbitrary) collector geometry. Figure 6.1 shows typical global hemispherical pyranometer installations on photovoltaic arrays to measure the plane of array irradiances.

Here, only the most commonly used models and their accuracy are described, as established in published validation studies. Note that for the most part, all of these approaches have been developed and validated using hourly average and in some cases lower time resolution, such as daily or monthly average solar radiation data. The validity of these model approaches and equations for higher time resolution, or subhourly data, is a matter of ongoing research.

TABLE 6.1
Solar Collector Energy-Harvesting Configurations

Configuration	Rationale	Approximate Enhancement Ratio with Respect to Horizontal Plane
Horizontal (0 tilt)	Low-latitude sites; flat roof installation; solar ponds	1.0
Latitude -15	Optimize for summer (winter) harvesting in Northern (Southern) hemisphere	1.2 (winter) to 0.8 (summer)
Latitude	Optimize annual average harvest	1.25–1.3
Latitude +15	Optimize for winter (summer) harvesting in Northern (Southern) Hemisphere	1.2 (summer) to 0.8 (winter)
Sun tracking (2-axis tracking)	Optimize collection of direct beam energy	1.3 to 1.5 (depending on climate)
East-west track (1 axis E-W)	Optimize morning and afternoon harvesting (track sun in azimuth)	1.15 to 1.2
North-south track (1 axis N-S)	Optimize harvesting through year midday (track sun in elevation)	1.2 to 1.3

FIGURE 6.1 (See color insert.) Global hemispherical pyranometers (left, silicon photo-diode; right, thermopile) measuring plane of array irradiance for photovoltaic performance testing.

6.2 ISOTROPIC MODELS

The angular distribution of the brightness, or radiance, of the sky dome is very variable. In clear conditions, the bright solar disk, circumsolar radiation, and distribution of energy in the sky dome (e.g., frequently observed "horizon brightening") range over several orders of magnitude and change throughout the day. Partly cloudy conditions are much more complex, including reflections from clouds, the ground to clouds, and then back to the ground, obscuration of the solar disk, and so on. Overcast conditions are slightly less complicated, with more uniform radiance distributions. Accordingly, models for estimating the total solar radiation available to an arbitrary collector configuration range from relatively simple to very complex.

The simplifying assumption of a relatively uniform (isotropic) radiation field can be used to produce simple equations for estimating the diffuse irradiance (ground-reflected radiation and the portion of the sky diffuse radiation) available to a nonhorizontal collector.

6.2.1 LIU AND JORDAN

Following the approach of Liu and Jordan [1] and Iqbal [2] for a surface tilted or inclined with respect to the horizontal, the direct beam incident on the inclined surface is modified by the cosine of the incidence angle θ_i of the beam on the plane. Since for a horizontal plane the incidence angle is the zenith angle z, the component of the DNI on the tilted plane I_t may be estimated from the DNI (I_b) using

$$I_t = I_b \cos(\theta_i)/\cos(z) \qquad (6.1)$$

where the incidence angle θ_i is calculated from Equation 1.12 or 1.13 in Chapter 1.

Next, the diffuse sky radiation on the inclined surface D_t and ground-reflected radiation on the surface R_t must be accounted for.

A simple approach is to apply the same incidence angle ratio $\cos(\theta_i)/\cos(z)$ to the diffuse horizontal irradiance (DHI), as was applied to the beam radiation in Equation 6.1, namely,

$$D_t = \text{DHI} \cos(\theta_i)/\cos(z) \qquad (6.2)$$

This approximation usually overestimates the diffuse sky radiation on the inclined plane. A somewhat better approach is to assume that the sky radiation is uniform over all directions and account for that portion "lost" by the portion of sky not "seen" by the inclined plane tilted at angle β from the horizontal. This leads to the following result:

$$\text{DTI} = 0.5\text{DHI}(1 - \cos \beta) \qquad (6.3)$$

Recalling Equation 1.22 from Chapter 1 for the total hemispherical radiation on an inclined plane (global tilted irradiance, GTI), we account for the direct radiation R reflected from the ground. For ground albedo or reflectivity of ρ, R consists of the reflected DNI, I_r, and DHI, D_r, impinging on the tilted surface. It can be shown that the following equations apply:

$$I_r = 0.5 \, \rho \, \text{DNI}[1 - \cos(\beta)] \qquad (6.4)$$

$$D_r = 0.5 \, \rho \, \text{DHI}[1 - \cos(\beta)] \qquad (6.5)$$

and

$$R = D_r + I_r \qquad (6.6)$$

For the global hemispherical irradiance on a tilted plane, GTI can be estimated as

$$GTI = I_t + DTI + R \qquad (6.7)$$

Comparing with Equation 1.22, I_t is either DNI $\cos(\theta_i)$ measured or modeled under clear skies or DNI $\cos(\theta_i)/\cos(z)$ where the DNI is computed from a general all-sky model or measured global horizontal data.

From the results of the Chapter 3 clear sky models for Barcelona, Spain, at latitude 41° N, on June 15 at 10 a.m., we can estimate the irradiance on a latitude tilted plane from

Direct beam radiation	DNI = 833 Wm²
Diffuse sky radiation	DHI = 139 Wm²
Total hemispherical radiation	GHI = 842 Wm²

Zenith angle, $z = 30°$, tilt angle, $\beta = 40°$, south facing; albedo (reflectivity) ρ of surrounding area, 0.20

From Equations 6.1 to 6.6,

$$
\begin{aligned}
I_t &= 833 \cos(40°)/\cos(30°) &&= 833(0.766/0.866) = 736 \\
DTI &= 139[1 - \cos(40°)] &&= 139 (1 - 0.766) \;\; = 33 \\
R &= [0.5(0.2)][1 - \cos(40°)](833 + 139) = (0.1)(0.234)(972) = 23
\end{aligned}
$$

and

$$GTI = I_t + DTI + R = 792 \text{ Wm}^2 \text{ on the inclined panel.}$$

6.3 ANISOTROPIC MODELS

The isotropic models described usually overestimate the available radiation for a tilted surface. Typically, there are brighter and dimmer portions of the sky dome, especially in the vicinity of the solar disk (circumsolar radiation) and the horizon (horizon brightening), even for clear skies. There is a long history of the development of anisotropic models in the work of Iqbal [3], and Muneer [4]. Temps and Coulson [5] developed a clear sky model that included circumsolar and horizon-brightening terms. This model was subsequently modified by Klucher [6] to include all sky conditions. The model modifies the basic isotropic Equation 6.3 with a circumsolar correction factor:

$$1 + \cos^2 \theta_i \sin^3 z \qquad (6.8)$$

and a horizon-brightening correction factor:

$$1 + \sin^3(\beta/2) \qquad (6.9)$$

The product of Equations 6.3, 6.8, and 6.9 then produces the clear sky diffuse irradiance on the tilted surface. Klucher's additional modification used a function of the ratio DHI/GHI to account for the sky condition:

$$F = 1 - (DHI/GHI)^3 \qquad (6.10)$$

Thus, Klucher's model computes

$$DTI = 0.5 \, DHI(1 - \cos \beta)(1 + F \cos^2 \theta_i \sin^3 z) \, [1 + F \sin^3(\beta/2)] \qquad (6.11)$$

Repeating the calculation for zenith angle, $z = 32.5°$, tilt angle, $\beta = 40°$, south facing; albedo (reflectivity) ρ of surrounding area 0.20 from our clear sky Barcelona example in Chapter 3, Section 3.1.4:

Direct beam radiation	$DNI = 833 \, Wm^2$
Diffuse sky radiation	$DHI = 139 \, Wm^2$
Total hemispherical radiation	$GHI = 842 \, Wm^2$

$\theta i = 38°$ (from Equation 1.13) and $z = 32.5°$; $F = 1 - (139/833)^3 = 0.995$

$$DTI = (0.5)(139)(1 - 0.766)[1 + (0.995)(0.621)(0.155)][1 + (0.995)(0.04)] = 18.5 \, Wm^2$$

which is compared with $DTI = 33 \, Wm^2$ from the isotropic model computed in the preceding section.

There are many other modifications and variations of the Temps/Coulson/Klucher anisotropic model, such as those of Hay [7], Willmott [8], Skartveit and Olseth [9], Reindl et al. [10], and Gueymard [11], with various differing functions to account for anisotropic sky conditions. All of these models perform about the same as the basic Klucher model. Mean bias errors are typically ±1%, and root mean square errors are about 10% [12]. The next section describes the anisotropic model of Richard Perez et al. [13, 14], which when compared with the models described previously as reported in the work of Kudish and Evseev [12], Loutzenheiser et al. [15], and Vartiainen [16], generally performs the best. For this reason, the Perez anisotropic model has become a popular tilt conversion model incorporated into most photovoltaic system design simulation software [17].

6.4 THE PEREZ ANISOTROPIC TILT CONVERSION MODEL

The Perez anisotropic model [13] produces the estimated diffuse irradiance on a tilted surface based on three modifier terms to the total hemispherical diffusion, DHI. The model equation is

$$DTI = DHI[(1 - F_1) \cos^2(\beta/2) + F_1(a_o/a_1) + F_2 \sin (\beta)] \qquad (6.12)$$

where $a_o = \cos \theta_i$, and $a_1 = \cos z$; thus, a_o/a_1 is the same ratio of cosine terms as in Equation 6.2. β is the tilt angle of the plane, and F_1 and F_2 are circumsolar and horizon

TABLE 6.2

Coefficients for Perez Model Fij Parameters

ε Bin	ε Low	ε High	F11	F12	F13	F21	F22	F23
1	1.000	1.065	−0.0083	0.5877	−0.0621	−0.0596	0.0721	−0.0220
2	1.065	1.230	0.1299	0.6826	−0.1514	−0.0189	0.0660	−0.0289
3	1.230	1.500	0.3297	0.4869	−0.2211	0.0554	−0.0640	−0.0261
4	1.500	1.950	0.5682	0.1875	−0.2951	0.1089	−0.1519	−0.0140
5	1.950	2.800	0.8730	−0.3920	−0.3626	0.2256	−0.4620	0.0012
6	2.800	4.500	1.1326	−1.2367	−0.4118	0.2878	−0.8230	0.0559
7	4.500	6.200	1.0624	−1.5999	−0.3589	0.2642	−1.1272	0.1311
8	6.200	...	0.6777	−0.3273	−0.2504	0.1561	−1.3765	0.2506

Source: Sandia National Laboratory. (1988). *Data from the Development and Verification of the Perez Diffuse Radiation Model.* Contractor Report Sand88-7030. Sandia National Laboratories, Albuqurque, NM. http://prod.sandia.gov/techlib/access-control.cgi/1988/887030.pdf.

brightening correction terms, respectively. The F terms are functions of the three parameters: the zenith angle Z, a clearness parameter ε, and a brightness parameter Δ. The clearness parameter e is derived from the DHI and DNI irradiance (note Z is in degrees):

$$\varepsilon = [(DHI + DNI)/DHI + 5.535 \cdot 10^{-6} \, Z^3]/[1 + 5.535 \cdot 10^{-6} \, Z^3] \qquad (6.13)$$

and

$$\Delta = m \, DHI/I_0 \qquad (6.14)$$

where m is the air mass, and I_0 is the extraterrestrial DNI or beam radiation.

In the original Perez article, the coefficients F_1 and F_2 were derived from six coefficients Fij, where i = 1 to 3 and j = 1 to 3. The Fij were selected based on eight brightness bins for ε with upper and lower bounds reported in a lookup table as in Table 6.2.

A simplified method of calculating the Fij parameters can be achieved by fitting polynomials to the coefficients to the ε bin number in Table 6.2 as shown in Figure 6.2. By then, using the epsilon bin boundaries to select the appropriate bin number, one can simply compute the appropriate Fij. Note the correlation coefficients for these fits all exceed 0.99.

For completeness, the polynomial expressions for the Fij coefficients as a function of BIN NUMBER x (from 1 to 8) are repeated here:

$$F11 = -0.0161x^3 + 0.1840x^2 - 0.3806x + 0.2324 \qquad (6.15)$$

$$F12 = 0.0134x^4 - 0.1938x^3 + 0.8410x^2 - 1.4018x + 1.3579 \qquad (6.16)$$

$$F13 = 0.0032x^3 - 0.0280x^2 - 0.0056x - 0.0385 \qquad (6.17)$$

$$F21 = -0.0048x^3 + 0.0536x^2 - 0.1049x + 0.0034 \qquad (6.18)$$

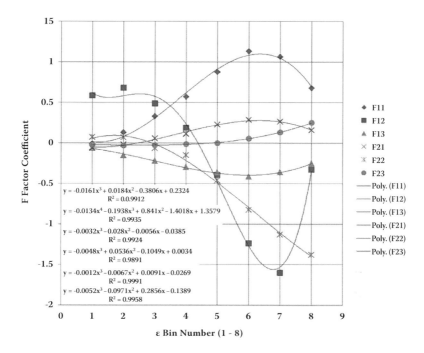

FIGURE 6.2   Plot of Perez Fij coefficients versus ε bin number from Table 6.2.

$$F22 = 0.0012x^3 - 0.0067x^2 + 0.0091x - 0.0269 \tag{6.19}$$

$$F23 = 0.0052x^3 - 0.0971x^2 + 0.2856x - 0.1389 \tag{6.20}$$

The F1 and F2 terms in Equation 6.12 are computed from (note zenith angle conversion to radians)

$$F1 = \max[0, F11 + F12\Delta + F13z(\pi/180)] \tag{6.21}$$

$$F2 = F21 + F22\Delta + F23z(\pi/180) \tag{6.22}$$

In the work of Perez et al. [14], coefficients Fij were computed for several U.S. and continental European sites, but in the end a comprehensive "aggregate" set of Fij was selected for the general model. Typical mean bias errors and root mean square errors for the model have been shown to be 5% and ±15%, respectively. Performance of the popular PVFORM photovoltaic system performance model, which uses the Perez tilt conversion model, has been evaluated to have similar accuracy: 5.9% mean bias, 12.5% root mean square errors [18]. These differences include additional components of uncertainty related to issues with the PVFORM model and the accuracy of the measured data used to compare with the PVFORM results.

6.4.1 COMPUTATIONAL EXAMPLE

From the results of the Chapter 3 clear sky models for Barcelona, Spain, at latitude $41°$ N on June 15 at 10 a.m., we can estimate the irradiance on a latitude tilted plane from:

Direct beam radiation   DNI = 833 Wm^2

Diffuse sky radiation   DHI = 139 Wm^2

Zenith angle, $z = 30°$, tilt angle, $\beta = 40°$, angle of incidence of direct beam on tilted south-facing surface = $38°$

$$a_0 = \cos(i) = \cos(38°) = 0.788$$

$$a_1 = \cos(z) = \cos(32.5°) = 0.843$$

$$a_0/a_i = 0.935$$

$$\varepsilon = [\{(139 + 833)/139\} + 5.535 \cdot 10^{-6}(32.5)^3]/[1 + 5.535 \cdot 10^{-6}(32.5)^3]$$
$$= (7.183)/[1.190] = 6.03$$

From Table 6.2, the epsilon bin is 7 ($4.5 < \varepsilon < 6.2$), and in Equations 6.15 to 6.20, $x = 7$:

F11 $= -0.0161x^3 + 0.1840x^2 - 0.3806x + 0.2324 = 1.062$

F12 $= 0.0134x^4 - 0.1938x^3 + 0.8410x^2 - 1.4018x + 1.3579 = -1.546$

F13 $= 0.0032x^3 - 0.0280x^2 - 0.0056x - 0.0385 = -0.352$

F21 $= -0.0048x^3 + 0.0536x^2 - 0.1049x + 0.0034 = 0.249$

F22 $= 0.0012x^3 - 0.0067x^2 + 0.0091x - 0.0269 = 0.120$

F23 $= 0.0052x^3 - 0.0971x^2 + 0.2856x - 0.1389 = -1.114$

$\Delta = m\ DHI/I_0 = (1.19)(139)/1366 = 0.121$

And from Equations 6.21 and 6.22,

F1 $= \max[0, F11 + F12\Delta + F13z(\pi/180)]$
$= \max[0,\{1.062 - 0.121(1.546) - 0.352(0.567)\}] = 0.675$

F2 $= F21 + F22\Delta + F23z(\pi/180) = 0.249 + 0.120(0.121) - 1.114(0.567) = -0.368$

DTI $= DHI[(1 - F_1)\cos^2(\beta/2) + F_1(a_0/a_i) + F_2\sin(\beta)]$
$= 139[(1 - 0.675)\cos^2(20°) + 0.675(0.935) - 0.368\sin(40°)] = 94.4$ Wm^2

Compare this value with the previously computed values of 18 Wm^2 for the Klucher model and 33 Wm^2 for the isotropic model.

Last, accounting for ground-reflected radiation R, assuming an isotropic reflectance model (Equations 6.4 and 6.5),

$$R = [0.5(0.2)][1 - \cos(40°)](139) = (0.1)(0.234)(139 + 833) = 23 \ Wm^2$$

The final hemispherical irradiance on the 40° tilted panel for this model is (a_0 is cosine of incidence angle for the DNI)

$$a_0(DNI) + DTI = (0.788)(833) + 94 + 23 = 753 \ Wm^2.$$

Part of the reason for the differences in the sky diffuse results, even for clear sky conditions (since we used the Bird clear sky model results in the calculations here) is the following:

- The isotropic model tends to underestimate sky diffuse radiation.
- The corrections applied in the Klucher model are relatively small for clear skies and small zenith angles.
- The ratio of DHI to DNI of 0.16 (16% diffuse) is indicative of relatively "hazy" clear skies; very clear skies have DHI/DNI of about 10%, or 0.1. The Perez epsilon and delta parameters reflect the influence of the additional sky diffuse under hazy clear sky conditions.

6.4.2 The Accuracy of the Perez Anisotropic Model

The accuracy of the Perez anisotropic model has been evaluated by both the original authors [13] and many other writers over the more than 20-year lifetime of the model. The original authors quoted "composite" uncertainty composed of average differences over five orientations (vertical, or 90° tilt, in cardinal directions—north, south, east, and west; and 45° south-facing tilt) for seven international locations (from table 8 in Perez, Ineichen, and Seals [13]). The overall composite random error ranged from 17% for 90° tilted north-facing surfaces to 4% for 45° tilted south-facing surfaces. The average absolute value of the random differences was 15 Wm^2. The systematic or bias errors were relatively small, from $-2 \ Wm^2$ to $+2 \ Wm^2$ on average, or about 0.5%.

An analysis of 12 models by Nooriana et al. [19] reported the Perez model as having systematic differences of -2% and random differences of 12% for south-facing inclined surfaces. The model was one of the best performers. The reference for this study was measured data on 45° south- and west-facing surfaces, and these were the only orientations studied. Random differences for the other models ranged from 10% to 55% and systematic differences from -1% to -32%.

A study in Japan [20] reported the probable error in estimating the error in the total irradiance on an inclined surface using the Perez model as ±50 Wm^2 each for both random and systematic error. A study in Singapore [21] reported a random error range for a 22° south-facing tilt from 4% to 9% for overcast skies and 7% for all sky conditions.

Note these uncertainties amount to about a factor of 2 larger than the measurement uncertainties described in Section 2.6.5 of Chapter 2 for pyranometer measurements. This is to be expected if the uncertainties are about the same and distributed equally among the measurement and model sources of error.

6.5   SUMMARY

Estimating the solar radiation on inclined surfaces such as those involved in the design of solar panels for energy harvesting is rather complex and difficult. Factors such as circumsolar aureole brightness and horizon brightening as well as variations in the foreground albedo of a tilted surface contribute to this complexity. Simple assumption of isotropic solar radiation fields can result in significant errors in estimation of the radiation available to inclined surfaces. Parameterization of the complex elements of solar aureole and horizon brightening may be highly site dependent, as shown in Perez et al. [14]. Uncertainties on the order of 10% to 15% have been achieved for anisotropic models. Especially under clear conditions, and in terms of absolute values of irradiance, the real level of accuracy for the estimation of irradiance on tilted surfaces may approach about twice the uncertainty in the measurement instruments. The relatively best-performing models still result in errors on the order of 50 Wm^2 or more in the total irradiance on an inclined surface under a wide variety of sky conditions. The design engineer should keep these considerations in mind when trying to convert GHI, DNI, or DHI into irradiance available to solar collectors in nonhorizontal configurations.

REFERENCES

1. Liu, B.Y.H., and R.C. Jordan. (1962). Daily insolation on surfaces tilted towards the equator. *Transactions of the ASHRAE*, Vol. 67, pp. 526–541.
2. Iqbal, M. (1980). Prediction of hourly diffuse solar radiation from measured hourly global radiation on a horizontal surface. *Solar Energy*, Vol. 24, pp. 491–503.
3. Iqbal, M. (1983). *An Introduction to Solar Radiation*. Academic Press, New York.
4. Muneer, T. (ed.). (2004). *Solar Radiation and Daylight Models*, 2nd ed. Elsevier Butterworth-Heinemann, Oxford, UK.
5. Temps, R.C., and K.L. Coulson (1977). Solar radiation incident upon slopes of different orientation. *Solar Energy*, Vol. 19, pp. 179–184.
6. Klucher, T.M. (1979). Evaluation of models to predict insolation on tilted surfaces, *Solar Energy*, Vol. 23, pp. 111–114.
7. Hay, J.E. (1979). *Study of Shortwave Radiation on Non-horizontal Surfaces*. Report 79-12. Atmospheric Environment Service, Downsview, Ontario, Canada.
8. Willmott, C.J. (1982). On the climatic optimization of the tilt and azimuth of flat-plate solar collectors. *Solar Energy*, Vol. 28, pp. 205–216.
9. Skartveit, A., and J.A. Olseth. (1986). Modeling slope irradiance at high latitudes. *Solar Energy*, Vol. 36, pp. 333–344.
10. Reindl, D.T., W.A. Beckman, and J.A. Duffie. (1990). Evaluation of hourly tilted surface radiation models. *Solar Energy*, Vol. 45, pp. 9–17.
11. Gueymard, C. (1987). An anisotropic solar irradiance for tilted surfaces and its comparison with selected engineering algorithms. *Solar Energy*, Vol. 38, pp. 367–386.

12. Kudish, A.I., and E.G. Evseev. (2009). Prediction of solar global radiation on a surface tilted to the south. *Proceedings SPIE*, Vol. 6652, Optical Modeling and Measurements for Solar Energy Systems. Society of Photo-optical Instrumentation Engineers, Bellingham, WA.

13. Perez, R., P. Ineichen, and R. Seals. (1990). Modeling daylight availability and irradiance components from direct and global irradiance. *Solar Energy*, Vol. 44, pp. 271–279.

14. Perez, R., R. Stewart, R. Seals, and T. Guerlin. (1988). *The Development and Verification of the Perez Diffuse Radiation Model*. Contractor Report Sand88-7030. Sandia National Laboratories, Albuquerque, NM. http://prod.sandia.gov/techlib/access-control.cgi/1988/887030.pdf. Accessed 21 July 2012.

15. Loutzenhiser, P.G., H. Manz, C. Felsmann, P.A. Strachan, T. Frank, and G.M. Maxwell. (2007). Empirical validation of models to compute solar irradiance on inclined surfaces for building energy simulation. *Solar Energy*, Vol. 81, pp. 254–267.

16. Vartiainen, E. (2000). A new approach to estimating the diffuse irradiance on inclined surfaces. *Renewable Energy*, Vol. 20, pp. 45–64.

17. Menicucci, D.F., and J.P. Fernandez. (1988). *User's Manual for PVFORM: A Photovoltaic System Simulation Program for Stand-Alone and Grid-Interactive Applications*. Sandia Report SAND85-0376. Sandia National Laboratories, Albuquerque, NM. http://prod.sandia.gov/techlib/access-control.cgi/1985/850376.pdf. Accessed 21 July 2012.

18. Perez, R., J. Doty, B. Bailey, and R. Stewart. (1994). Experimental evaluation of a photovoltaic simulation program. *Solar Energy*, Vol. 52, 4, pp. 359–365.

19. Nooriana, A.M., I. Moradib, and G.A. Kamali. (2008). Evaluation of 12 models to estimate hourly diffuse irradiation on inclined surfaces. *Renewable Energy*, Vol. 33, pp. 1406–1412.

20. Unozawa, H., K. Otani, and K. Kurokawa. (2001). A simplified estimating model for in-plane irradiation using minute horizontal irradiation. *Solar Energy Materials and Solar Cells*, Vol. 67, pp. 611–619.

21. Li, D.H.W., and G.H.W. Cheung (2005). Study of models for predicting the diffuse irradiance on inclined surfaces. *Applied Energy*, Vol. 81, pp. 170–186.

7 Introduction to Modeling Spectral Distributions

Science is spectral analysis. Art is light synthesis.

—Karl Kraus, 1912

7.1 THE SPECTRAL ATMOSPHERIC FILTER

As described in previous chapters, the atmosphere acts as a continuously variable filter affecting the solar radiation propagating to Earth's surface. Atmospheric gases, aerosols and particles, water vapor and droplets (clouds), and various pollutants modify the distribution of solar energy with respect to wavelength, or "spectral distributions" as we will refer to them. The result is a wide range of variation in the spectral distribution of natural sunlight with respect to local conditions, time of day, and so on. In this chapter, we provide a brief summary of the importance of spectral distributions to renewable energy applications, an overview of the principles behind various complex and simple spectral models, and the equations behind one of the most simple and straightforward spectral models. Note that the following sections except for Sections 7.3 and 7.8 address spectral models under clear (cloudless) skies only.

7.2 RENEWABLE ENERGY APPLICATIONS FOR SPECTRAL DATA AND MODELS

There is a wide variety of materials and systems, including the human vision system, plant photosynthesis response system, and photovoltaic (PV) material combinations and systems, that have limited and different spectral response functions. The combination of the variability in natural spectral distributions and spectral response functions presents significant challenges in designing, evaluating, and predicting the performance of renewable energy systems, including daylighting, PV systems, and optical properties of materials in general, especially in outdoor applications.

Spectral measurements rely on the collection and sorting of optical radiation into the spectral, or wavelength-by-wavelength, components. The instrumentation relies on dispersion elements such as prisms or, more commonly, diffraction gratings, and detectors with the sensitivity to produce adequate signals proportional to stimuli that are from five to six orders of magnitude smaller than the total integrated broadband signals. The development of arrays of solid-state and charge-coupled device (CCD) detectors in conjunction with spectral dispersion elements has lowered the cost of the measurement technology somewhat, but calibration and operations of these measurement devices is still expensive. In addition, massive amounts of data under a variety of conditions may be acquired quite quickly. Organizing such data streams, along with the "luck of the draw" in terms of prevailing conditions, is also problematic. This means that measurements, especially in the field rather than the laboratory, are expensive and labor intensive [1]. Just as with broadband solar radiation modeling, the modeling of spectral distributions may provide useful information in the absence of actual measured data. We briefly describe models for generating spectral irradiance distributions for use in solar renewable applications.

The limited spectral response range of PV materials has already been mentioned several times and was illustrated in Figure 2.6 of Chapter 2. At the top of that figure, the spectral response range for a variety of PV materials is shown. Also, the spectral range for visible (photopic) and photosynthetically active radiation (PAR, 400 to 700 nm) is shown. Plants capture photons in the PAR region to convert carbon dioxide into organic compounds for growth. The human photopic response region (380 to 830 nm) of course relates to our ability to discern colors and human vision in general. The ultraviolet (UV) spectral regions are important for ozone formation in the stratosphere and sunburn and cancer in humans but also are key drivers of material degradation, both in solar energy systems and in general. The infrared regions of the spectrum are important in the "greenhouse" effect, both in real greenhouses for plant growth and in our atmosphere. But, beyond those applications, the infrared region is important for the design of energy-efficient glass materials for buildings, especially in large structures. In addition, the spectral optical properties of materials can be key to designing and computing building heating and air conditioning loads. Thus, spectral information, especially since it may be highly variable, can be valuable information for all of these applications.

7.3   COMPLEX SPECTRAL MODELS

Modeling radiative transfer through an absorbing, scattering, and emitting medium has a long history [2]. Research into physical and optical properties of Earth's atmosphere and their effect on the propagation of solar radiation has resulted in the definition of standard atmospheres. These include the United States Standard Atmosphere (USSA) of 1966, revised in 1976 [3], and supplemental atmospheric models [4]. Similarly, a wide variety of radiative transfer models of differing complexity has evolved.

One means of computing radiative transfer is Monte Carlo modeling. Interactions of individual photons with the physical properties of the media are modeled using random processes. For accuracy, a large number of photons and iterative computations is needed at each wavelength. The BRITE model of Blättner et al. [5] is an example. Radiative transfer through an atmospheric path depends on quantum properties of the atmospheric constituents. High-resolution models using these quantum properties are called line-by-line or LBL models [6–8]. An example is Fast Atmospheric Signature Code (FASCODE), developed by the Air Force Geophysics Laboratory (AFGL) [8]. These models are for narrow-bandwidth regions and require significant computational resources and storage space. LBL models access databases, such as HITRAN (high-resolution transmission) [9], consisting of quantum parameters for many molecular species (more than 1 million spectral lines). LBL models are too complex and specialized for discussion here. Less-complex "band" models are simplified LBL models as described in Fenn et al. [6]. Band models represent groups of absorption lines as transmittance functions of parameters such as absorber concentration, pressure, and absorption coefficients.

7.3.1 MODTRAN AND LOWTRAN

MODTRAN® (moderate resolution) and LOWTRAN® (low resolution), developed by AFGL, are popular, commercially available band models [10]. They are also complex; MODTRAN is comprised of over 50,000 lines of FORTRAN program code and over 200 subroutines. "Low" resolution corresponds to 20 wavenumbers, usually quoted as "reciprocal centimeters" (cm^{-1}), denoting the frequency of the photons. For example, a resolution of 20 cm^{-1} converts to 0.2 nm at 300 nm and to 32 nm at 4000 nm in wavelength terms. "Moderate" resolution corresponds to 2 cm^{-1} and to 0.02 nm at 300 nm and 3 nm at 4000 nm. These models can address complex scenarios, including clouds, fog, smoke, many choices of standard and user-defined aerosol properties, atmospheric structure for up to 33 different layers, and a selection of different extraterrestrial spectra. They are primarily designed to compute atmospheric transmittance between two points on or above Earth's surface for military and remote sensing applications. The many combinations of input parameters and their interaction require a great deal of understanding by the user. Interpretation of the sometimes-huge output data files (approaching 100 megabytes in size at times) is daunting as well.

7.3.2 LibRadtran and Other Complex Spectral Models

A second example from the family of complex radiative transfer codes is the Library for Radiative Transfer, or libRadtran, literally a library of radiative transfer routines and programs [11]. The central program of the libRadtran package was originally designed to calculate spectral irradiance in the UV and visible parts of the spectrum. Over the years, numerous extensions and improvements have included the full solar and thermal spectrum, currently from 120 nm to 100 μm. LibRadtran complexity is similar to that of MODTRAN/LOWTRAN but was developed in the academic environment. The model provides a variety of options to set up and modify an atmosphere with molecules, aerosol particles, water and ice clouds, and a surface as lower boundary. LibRadtran includes a selection of about 10 different radiative transfer equation solvers, fully transparent to the user. The LibRadtran manual contains very detailed and useful descriptions of the complex physics, data, and algorithms required for high-resolution, physics-based spectral models [11]. An extensive list of radiative transfer codes and algorithms, some of which are freely available for download, is presented online (http://en.wikipedia.org/wiki/Atmospheric_radiative_transfer_codes, accessed 21 July 2012).

7.4 STANDARD SPECTRAL DISTRIBUTIONS

Atmospheric conditions such as water vapor content, dust and aerosols, smoke from biomass burning, and pollution are highly variable. Therefore, there is a need for benchmark or reference spectra, especially in PV energy conversion, to serve as a common frame of reference for comparison for various materials under the same spectral distribution. The fundamental starting point for all spectral modeling of terrestrial solar spectra (i.e., spectra somewhere within Earth's atmosphere) is the spectral distribution of the extraterrestrial solar radiation at the top of the atmosphere or so-called air mass zero (AM0) spectrum. There is a wide variety of available measured extraterrestrial spectral distributions, collected from different platforms over a wide period of time by sensors of varying quality. This has resulted in the need for a single consensus standard extraterrestrial spectrum for aerospace applications.

7.4.1 REFERENCE AM0 EXTRATERRESTRIAL SPECTRA

The earliest attempts at estimating the extraterrestrial or AM0 spectral distribution were based on extrapolating measured spectral data measured at high altitudes (mountaintops) to the top of the atmosphere and correcting for atmospheric effects. Later, high-altitude balloon and aircraft flights with mounted instruments were flown as high into the upper atmosphere as possible, as described in Drummond and Thekaekara [12]. With the coming of the space age, a wide variety of spacecraft instrumentation (not always in a coordinated fashion) was to sent into Earth's orbit to measure various regions of the solar spectrum of scientific interest. The results of these various missions were used by many authors to assemble extraterrestrial solar spectra in a piecemeal fashion. Examples include the work by Wherli [13] and Neckel and Labs [14]. An extensive overview and history was provided by Gueymard [15]. Not until 2003 was a dedicated, wideband, moderate-resolution spectroradiometer (in addition to a broadband solar irradiance monitor) deployed on the Solar Radiation and Climate Experiment (SORCE) satellite. This instrument measures the solar spectrum from 200 nm in the deep UV to 2400 nm in the near infrared [16, 17]. With the advent of extremely fast and powerful computers, computational astrophysics were used to compute the solar (and other stellar) spectra from first principles, as described by Kurucz [18].

In the 1980s, the American Society for Testing and Materials (ASTM) developed an AM0 reference spectrum (ASTM E490) for use by the aerospace community [19] based on the early work of Drummond and Thekaekara. Since 2006, the ASTM E490 Air Mass Zero solar spectral irradiance is based on data from satellites, space shuttle missions, high-altitude aircraft, rocket soundings, ground-based solar telescopes, and modeled spectral irradiance. The integrated spectral irradiance has been made to conform to the value of the solar constant accepted by the space community, which is 1366.1 Wm². The complexity of the assemblage of these spectra is illustrated by a combination of sources used. The data were assembled from two different instruments on the Upper Atmosphere Research Satellite (UARS), the Solar Ultraviolet Spectral Irradiance Monitor (SUSIM), and the Solar Stellar Irradiance Comparison Experiment (SOLSTICE), reported by Woods et al. [20]; data from the McMath Solar Telescope at Kitt Peak, Arizona; the high-resolution solar atlas computed by Kurucz; and irradiance versus wavelength fits reported by Smith and Gottlieb [21]. The composite spectral irradiance data were then scaled to force the integrated total irradiance equal to the solar constant.

There are other AM0 solar spectra that have been derived since 1970 [8]. For historical interest, there is the 1973 "Thekaekara" spectrum, which is often quoted. Several of these spectra are available as user-selectable starting points in the MODTRAN (five different spectra), libRadtran, and SMARTS (Simple Model of the Atmospheric Radiative Transfer of Sunshine) (nine different spectra) models for atmospheric spectral transmission of sunlight. The MODTRAN AM0 spectra have very high spectral (wavelength) resolution, with approximately 50,000 data points at 1-wavenumber (1-cm^{-1}) resolution. There are also extremely high-resolution spectra (0.005 nm or 5 angstroms or better) available at the following Web and FTP download sites:

- http://www.mmnt.net/db/00/ftp.hs.uni-hamburg.de/pub/outgoing/FTS-Atlas accessed 26 Oct 2012.
  Described in "Spectral Atlas of Solar Absolute Disk-Averaged and Disk Center Intensity from 3290 to 12510 A," *Solar Physics,* 184(2), 421, February 1999 (unpublished data from Brault and Neckel, archived in 1987, and available from Hamburg Observatory anonymous FTP site). This atlas contains over 1 million data points broken up into 1000-Å wavelength band files.
- http://www.noao.edu/image_gallery/html/im0600.html (accessed 21 July 2012).
  Here, one can find a high-resolution pictorial version of the spectrum of the sun, created from a digital atlas observed with the Fourier transform spectrometer at the McMath-Pierce Solar Facility at Kitt Peak National Observatory, Sunspot, Arizona ("Solar Flux Atlas from 296 to 1300 nm," Robert L. Kurucz, Ingemar Furenlid, James Brault, and Larry Testerman, *National Solar Observatory Atlas No. 1,* June 1984). The "colors" of the spectrum are shown with the dark (Fraunhofer) absorption lines shown.
- http://bass2000.obspm.fr/solar_spect.php accessed 26 Oct 2012.
  Provides subangstrom resolution spectral data in a requested band in graphical and ASCII formats from 300 to 5400 nm.

7.4.2 REFERENCE TERRESTRIAL SPECTRA FOR RENEWABLE ENERGY APPLICATIONS

As the PV industry evolved, a consistent reference spectrum for comparing performance of various PV technologies was needed. ASTM standard E1036 on standard reporting conditions (SRC) for PV device performance prescribes a device temperature of 25°C, a total irradiance on the plane of the device of 1.0 kWm^{-2}, and a specified standard spectral distribution [22]. The specified standard reference spectra are presently described in ASTM G173-03 [23]. Older versions of the reference spectra, first developed in 1982, were based on older radiative transfer calculations that could not be easily reproduced [24]. Therefore, ASTM worked to develop G173 as a well-documented, easily accessible model for generating the reference spectra. Version 2.9.2 of the Gueymard's SMARTS model [25] was selected as appropriate. The model itself is integrated into the revised standard as an adjunct to the standard (part number ADJG173CD available from ASTM).

The principal differences between the old and new reference spectra, other than the operational model used to generate the spectral data, are the AM0 spectra available, and the default aerosol optical depth (AOD) values assumed. The SMARTS model provides a user choice of AM0 spectrum to use, derived from the MODTRAN AM0 spectra. The MODTRAN AM0 spectra have been resampled to 0.5-nm resolution between 280 and 400 nm, 1.0-nm resolution from 400 to 1700 nm, and 5-nm resolution between 1700 and 4000 nm, with a transition wavelength at 1702 nm. The resampling computed the equivalent average energy in the SMARTS bands from the higher-resolution MODTRAN data within the SMARTS band by integration over the bandwidth and division by the bandwidth. In addition, AM0 spectra developed by Gueymard from more up-to-date satellite data are provided [15].

The updated reference standard spectra use a default AOD of 0.084 (at 500 nm) versus the previous significantly higher AOD of 0.27 in the older standard. This change was made as the result of an analysis of conditions when irradiances on a flat plate met SRC for irradiance of 1000 Wm^{-2} [26]. The change resulted in the integrated hemispherical spectral irradiance on a flat plate of 1000 Wm^{-2}. The older standard and model produced lower total irradiance integrals that required artificial adjustment factors to produce the 1.0-kWm^{-2} integrated spectra. The updated standard also increased the direct beam spectrum, and beam total irradiance, to 900 Wm^{-2}, a level comparable with proposed concentrator rating methods such as ASTM E2527 [27].

The International Electrotechnical Commission (IEC) has produced a standard reference hemispherical spectral distribution (IEC 90804-03) [28] that is based on, but slightly modified from, the ASTM G173 spectrum. That standard also utilizes the SMARTS version 2.9.2 model to generate the spectra. The IEC standard includes only the hemispherical spectral distribution and not a direct beam spectrum. For a detailed discussion of the (minor) differences, see Myers [29].

7.4.3 THE INTERNATIONAL COMMISSION ON ILLUMINATION

The CIE, or International Commission on Illumination, derives the abbreviation from the French name of the organization, *Commission Internationale de l'Éclairage*, is the international authority on light, illumination, color, and color spaces. It was established in 1913 as a successor to the *Commission Internationale de Photométrie* and is today based in Vienna, Austria. Many CIE documents and reports are internationally recognized standards, primarily in the areas of radiometry, vision, color, illumination, and optical properties of materials. There are several CIE documents relating to spectral distribution of daylight and artificial illuminants.

The CIE publishes well-known standard illuminants. Each of these is known by a letter or by a letter-number combination. Illuminants A, B, and C were introduced in 1931, with the intention of respectively representing average "incandescent light," "direct sunlight," and "average daylight" [30]. Illuminants D represent phases of daylight, illuminant E is an equal-energy illuminant, while illuminants F represent fluorescent lamps of various composition. There are instructions on how to experimentally evaluate light sources ("standard sources") corresponding to the daylight illuminants [31].

At present, no artificial source is recommended to realize CIE standard illuminants for daylight. It is hoped that new developments in light sources and filters will eventually offer a sufficient basis for a CIE recommendation [32].

7.5 CIE SPECTRAL MODEL—ILLUMINANT D65 AND DAYLIGHT

Architects and illumination engineers rely on the CIE for many technical aspects of their design needs and with respect to replacing artificial lighting with daylighting. The CIE illuminant D65 represents a "phase of daylight" with correlated color temperature (CCT) of 6500 K. The color temperature of a light source is the temperature of an ideal black-body radiator that radiates light of comparable hue to that of the light source. Color temperature is stated in the unit of absolute temperature, the kelvin (K). The CCT T_c is a specification related to the color appearance of the light emitted by a source, relating its color to the apparent color of a black-body radiator of the same temperature. The designations of variations on the D65 spectra based on different CCTs are generally reported as DXX, where XX are the two most significant digits of the CCT in thousands of kelvin. CIE recommends calculating the *relative* spectral distributions of other phases of daylight at different CCTs based on the CCT of the phase of daylight desired. Two steps are specified (see [33]):

Chromaticity: The (1931) x, y chromaticity coordinates (a locus within CIE-defined "color space") of the daylight with CCT T_c in kelvin to be calculated must satisfy

$$y_D = -3.0x_D^2 + 2.870x_D - 0.275 \text{ with } 0.250 < x_D < 0.380 \tag{7.1}$$

where the CCT T_c of daylight is used to compute x_D by two formulas, depending on the range of T_c:

$$4000 \text{ K} < T_c \le 7000 \text{ K: } x_D = -4.607(10^9/T_c^3) + 2.9678(10^6/T_c^2)$$
$$+ 0.09911(10^3/T_c) + 0.244063 \tag{7.2}$$

$$7000 \text{ K} < Tc \le 25000 \text{ K: } x_D = -2.0064(10^9/T_c^3) + 1.9018(10^6/T_c^2)$$
$$+ 0.24748(10^3/T_c) + 0.23704 \tag{7.3}$$

Relative Spectral Distribution: The relative (with respect to the defined reference spectrum D65) spectral irradiance distribution for the source with CCT of T_c is then calculated from

$$S(l) = S_0(l) + M_1S_1(l) + M_2S_2(l) \tag{7.4}$$

where M_1 and M_2 modifier functions are computed from x_D and y_D using

$$M_1 = (-1.3515 - 1.7703x_D + 5.9114y_D)/(0.0241 + 0.2562x_D - 0.7341y_D) \tag{7.5}$$

$$M_2 = (-0.030 - 31.442x_D + 30.0717y_D)/(0.0241 + 0.2562x_D - 0.7341y_D) \tag{7.6}$$

For this calculation, the D65 spectrum and tabulated values of $S_0(l)$, $S_1(l)$, and $S_2(l)$ are given in the table in Appendix C.

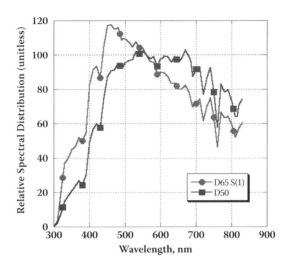

FIGURE 7.1 Relative amplitudes of CIE D65 reference spectral distribution (circles) and 5000 K CCT "daylight" spectrum (squares; D50) derived using the CIE spectral model. (Adapted from public domain CIE data. Data from http://www.cie.co.at/publ/abst/datables15_2004/std65.txt.)

For example, using a CCT of 5000 K, the computed values of x_D, y_D, M_1, and M_2 would be

$$x_D = 0.346,\ y_D = 0.359,\ M_1 = -1.039,\ M_2 = 0.363.$$

Used in conjunction with the tabulated values for S_0, S_1, and S_2 in Appendix C, Figure 7.1 shows the relative spectral distribution of the CCT 5000 K "daylight" compared with the CIE D65 illuminant spectral distribution. Note the unitless nature of the amplitudes in Figure 7.1. Also, the wavelength range of the CIE D65 illuminants cover only the photopic response spectral region (300 to 830 nm) as these spectra are used mainly with respect to their impact on human visual response.

Comparing Figure 7.1 with Figures 2.3 to 2.6 of Chapter 2, the resemblance of the spectral shape of the D65 spectrum to the direct beam spectral distribution is apparent. As the CCT of the daylight is decreased, the spectrum shifts to having more energy in the longer-wavelength bands (i.e., becomes "redder"). Attempts are made to simulate the D65 (and variant spectra) with various artificial light sources to measure and evaluate the color and appearance of materials in direct sunlight. However, as mentioned, no specific artificial source or source/filter combination is recommended as the "reference standard" source for D65 or other phases of daylight. There are standards for the evaluation of the spectral quality of daylight (solar) simulators, from both CIE [31] and ASTM [34].

7.6   BIRD CLEAR SKY SPECTRAL MODEL SPCTRL2

For more general applications, including PV and optical properties of materials applications, Bird and Riordan [35, 36] and Bird and Hulstrom [37] applied many of the same principles used in the development of the broadband clear sky irradiance model described in Chapter 3. The model called SPCTRL2 by Bird is also variously referred to in the literature as SPCTRAL2 or SPECTRL2 and is based on Equation 3.1 of Chapter 3 written in terms of functions of wavelength:

$$Ib(\lambda) = Rc\ Io(\lambda)\ Tr(\lambda)\ Ta(\lambda)\ Tg(\lambda)\ To(\lambda)\ Tw(\lambda) \tag{7.7}$$

where Io is the AM0 spectrum; Tr, Ta, Tg, To, and Tw are transmittance functions for Rayleigh scattering, aerosols, uniformly mixed atmospheric gases, ozone, and water vapor (all as a function of wavelength λ), respectively; Rc is the Earth–Sun radius vector correction; and $Ib(\lambda)$ is the resulting direct beam spectrum. Bird used the Wehrli AM0 spectrum [13] as the starting point for his model. Bird selected 122 wavelengths at approximately 10-nm intervals from 305 to 4000 nm for the spectral calculations. The data for the Wehrli AM0 spectrum and the absorption coefficients for each of the model components in Equation 7.7 are listed in Table D.1 in Appendix D.

Rc can be computed from Equation 1.1 and 1.2 of Chapter 1. The Rayleigh and mixed gas absorption are functions of the air mass alone. The ozone, water vapor, and aerosol absorption depend on the concentration of these components in the atmosphere. The ozone amount can also be estimated from Heuklon's model, Equations 3.17–3.19 of Chapter 3. Water vapor can be estimated from relative humidity [38] or dew point [39]. Aerosol absorption is dependent on the AOD, usually a difficult parameter to obtain. However, as described in Chapter 3, the NASA AERONET network can provide information on the climatic values and variations in all of these parameters on a selected basis (see Table 3.1).

7.6.1 Spectral Transmission Functions

The form of the transmission functions for the SPCTRL2 model take the following forms (variables of the form $a_{x\lambda}$ are wavelength-dependent absorption coefficients for parameter x, and exp(x) is the exponential function e^x):

$$Tr(\lambda) = \exp\{-M'/[\lambda^4(115.6406 - 1.335/\lambda^2)]\} \tag{7.8}$$

and M' is the pressure-corrected air mass,

$$Tw(\lambda) = \exp[-0.2385a_w(\lambda)\ W\ M/(1 + 20.07a_w(\lambda)\ W\ M)^{0.45}] \tag{7.9}$$

where w is the estimated total precipitable water vapor amount (atm-cm), $a_w(\lambda)$ is the water vapor absorption coefficient as a function of wavelength, and M is the air mass (Note: *not* pressure corrected);

$$To(\lambda) = \exp(-a_o(\lambda)\ O_3\ M_o) \tag{7.10}$$

where O_3 is the ozone amount, Mo is the ozone air mass, and $a_o(\lambda)$ is the ozone absorption coefficient as a function of wavelength.

$$Mo = (1 + h_0/6370)/\cos(z)^2 + 2\ h_0/6370)^{0.5} \tag{7.11}$$

where h_0 is the altitude of maximum O_3 concentrations, about 22 km.

$$Tg(\lambda) = \exp[-1.41a_g(\lambda) \, M'/(1 + 118.93a_g(\lambda) \, M')^{0.45}] \qquad (7.12)$$

and $a_g(\lambda)$ is the mixed gas absorption coefficient as a function of wavelength.

For aerosol transmittance, the Angstrom formulation

$$Ta(\lambda) = \beta\lambda^{-\alpha} \qquad (7.13)$$

is used with $\alpha = 1.14$ for rural aerosol conditions, and β is the "Angstrom turbidity coefficient" representing the "clarity" of the atmosphere, ranging from pristine conditions with $\beta = 0$ to very hazy conditions with $\beta > 0.5$, as described in Chapter 3.

7.6.2 DIFFUSE SPECTRAL IRRADIANCE ON A HORIZONTAL SURFACE

The diffuse irradiance on a horizontal surface is divided into three components: the Rayleigh scattering component $Ir(\lambda)$, the aerosol scattering component $Ia(\lambda)$, and the component that accounts for multiple reflections of irradiance between the ground and the air $Ig(\lambda)$.

The total scattered irradiance $I_{s\lambda}$ is equal to $Ir(l) + Ig(l) + I_{g\lambda}$. Assuming the Rayleigh and aerosol scattering are independent of each other, the following expressions were developed by Bird:

$$Ir(\lambda) = 0.5I_0(\lambda) \, Rc \, \cos(z) \, To(\lambda) \, Tg(\lambda) \, Tw(\lambda) \, Ta(\lambda) \, (1 - Tr(\lambda)^{0.95}) \qquad (7.14)$$

$$Ia(\lambda) = H_0(\lambda) \, Rc \, \cos(z) \, To(\lambda) \, Tg(\lambda) \, Tw(\lambda) \, Taa(\lambda) \, Tr(\lambda)^{1.5} \, (1 - Tas(\lambda)) \, Fs \, Cs \qquad (7.15)$$

$$Ig(\lambda) = [I_0(\lambda) \cos(z) + Ir(\lambda) + Ia(\lambda)] \, rs(\lambda) \, rg(\lambda) \, Cs/(1 - rs(\lambda) \, rg(\lambda)) \qquad (7.16)$$

It is assumed that half of the Rayleigh scatter is downward regardless of the zenith angle of the sun, and that a fraction Fs of the aerosol scatter is downward and can be a function of the solar zenith angle. The transmittance terms $Taa(\lambda)$ and $Tas(\lambda)$ are for aerosol absorption and aerosol scattering, respectively. The terms and functions are defined as

$$rs(\lambda) = To'(\lambda) \, Tw'(\lambda) \, Taa'(\lambda)[0.5(1 - Tr'(\lambda)) + 0.1906Tr'(\lambda)(1 - Tas'(\lambda))] \qquad (7.17)$$

$$Tas(\lambda). = \exp(-\omega(\lambda) \, Ta(\lambda) \, M) \qquad (7.18)$$

where

$$\omega(\lambda) = 0.945\exp\{-0.095 \, [\ln(\lambda/0.4)]^2\} \qquad (7.19)$$

$$Taa(\lambda). = \exp[-(1 - \omega(\lambda)) \, \tau a(\lambda) \, M] \qquad (7.20)$$

$$Fs = 1 - 0.5\exp[(-1.8341 + 0.17599 \cos(z)) \cos(z)] \tag{7.21}$$

$$Cs = (\lambda + 0.55)^{1.8} \text{ for } \lambda \leq 0.45 \text{ }\mu m = 1.0 \text{ for } \lambda > 0.45 \text{ }\mu m \tag{7.22}$$

The parameter $rg(\lambda)$ is the ground albedo as a function of wavelength, $rs(\lambda)$ is the sky reflectivity, and the primed transmittance terms are the regular atmospheric transmittance terms evaluated at $M = 1.8$. $\omega(\lambda)$ is the aerosol single-scattering albedo as a function of wavelength. The constant terms in Equations 7.19 and 7.21 are the result of evaluating certain equations (Equations 3.12 to 3.16) in Bird and Riordan [36] with fixed values of the single-scattering albedo at 400 nm (0.65), aerosol asymmetry factor (0.65), and wavelength variation factor (0.095).

7.6.3 Diffuse Spectral Irradiance on a Tilted Surface

Bird and Riordan based the development of the diffuse spectral irradiance on a tilted surface from the Hay tilt algorithm (mentioned in Section 6.3).

$$It(\lambda) = Id(\lambda) \cos(\theta) + Is(\lambda) [Id(\lambda) \cos(\theta)/(Io(\lambda) Rc \cos(z)) + 0.5 (1 + \cos(s))$$

$$(1 - Id(\lambda)/(Io(\lambda) Rc))] + 0.5 [Id(\lambda) \cos(z) + Is(\lambda)] rg(\lambda) (1 - \cos (\theta)) \tag{7.23}$$

where θ is the tilt angle for the surface.

A FORTRAN listing of the SPCTRL2 model is provided in Appendix D. The program requires input values for the zenith angle, surface tilt angle, angle of incidence, day of the year, ozone amount, water vapor amount, surface pressure, ground albedo, and AOD at 500 nm. Representative water vapor and AOD values are available from the NASA AERONET Web site, as related in Chapter 3.

7.7 GUEYMARD CLEAR SKY SPECTRAL MODEL SMARTS

Highly detailed descriptions of the SMARTS model are provided in Gueymard and others [24, 25]. The approach is essentially a more sophisticated application of parameterized models of Bird and others (as in Equation 7.7), but using transmittance functions with additional terms for trace gases.

A characteristic of this model is its versatility: A large number of applications in various disciplines are possible. This is achieved by providing a number of options in addition to the core calculations. Other features of the model are as follows: (1) It uses accurate and regularly updated spectral transmittance functions; (2) it provides improved spectral resolution over existing transmittance models; (3) it produces spectral irradiances comparable to MODTRAN predictions with far simpler inputs; and (4) its predictions can be easily and directly compared to spectroradiometric measurements using built-in functions. Here, we present a succinct summary of the model elements and their derivation.

The philosophy behind the latest model developments is to parameterize the band model transmittance functions used by MODTRAN for water vapor (the strongest absorber in the infrared with very complex absorption features), but at a lower resolution of 0.5 nm in the UV less than 400 nm, 1 nm between 400 and 1700 nm, and 5 nm between 1700 and 4000 nm. Recent spectroscopic data were used to parameterize the temperature-dependent absorption by other gases. Rayleigh scattering was parameterized as a function of wavelength and pressure based on recent depolarization data. Aerosol transmittance was parameterized using Ångström's law and band-integrated values of Ångström's turbidity coefficients for a variety of aerosol models. These parameterized transmittance functions were used to obtain the direct beam irradiance from

$$E(\lambda) = Io(\lambda) \ Tr(\lambda) \ To(\lambda) \ Tmg(\lambda) \ Ttg(\lambda) \ Tw(\lambda) \ Ta(\lambda) \tag{7.24}$$

at each wavelength (λ, nm), where E is the terrestrial spectral irradiance, Io is the extraterrestrial spectral irradiance, and the spectral transmittances are for Rayleigh scattering (Tr), ozone absorption (To), mixed gas absorption (Tg), trace gas absorption (Ttg), water vapor absorption (Tw), and aerosol extinction (Ta). To define the amount of variable gases, five predefined pollution levels are selectable (pristine/exceptionally clean, standard/clean, light pollution, moderate pollution, heavy pollution), as well as user-specifiable mixing-layer pollutant concentrations. Version 2.9.2 of this model is the generating model for the ASTM International and IEC standard reference spectra for PV applications (ASTM G173, IEC 60904-3), UV radiation weathering applications (ASTM G177), and fenestration applications (ASTM G197).

Although model principles and equations are similar to the Bird SPCTRL2 mode, they are much more detailed and complex than those of the Bird SPCTRL2 model. The FORTRAN code for SPCTRL2 has about 200 lines or so of functional code, while SMARTS2 (also in FORTRAN) has about 2000 lines of code. The SMARTS2 model is available for free download through the National Renewable Energy (NREL) Renewable Resource Data Center (RRedC) (http://www.nrel.gov/rredc/smarts). The NREL download package contains the source and executable code, user's guide, and an Excel-based (for Windows operating system machines) user interface to assemble the "card deck" input file for model runs. The card deck structure of lines of data for input such as comment/title text, selection of the reference atmosphere, aerosol profiles, constituent values and concentrations, solar geometry, and so on, usually among about 20 to 30 lines of input values with one to three parameters per line. The input files can be assembled using a test editor or the Excel user interface developed for the Microsoft operating system platforms.

Users of the model have developed other interfaces for operating the model, including Labview for Apple operating systems. The developer of the model, Dr. Gueymard, provides updates and tools through his Web site (http://solarconsultingservices.com).

Several authors have examined the performance of the SPCTRL2 and SMARTS2, as well as more complex spectral models with respect to each other and measured spectra [39–42]. In general, the accuracy of the SPCTRL2 and SMARTS2 models is such that they compare within the uncertainty envelope of spectral irradiance measurements (on the order of about 5%).

7.8   SPCTRL2 AND SMARTS FOR ASTM STANDARD REFERENCE CONDITIONS

To show the relative differences between the SPCTRL2 and SMARTS readily available simple spectral models, we show the results of running both models using the atmospheric conditions prescribed in the present ASTM G173 reference spectrum, namely,

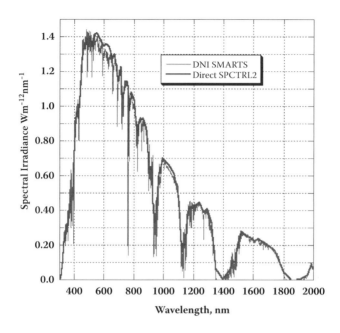

FIGURE 7.2 SPCTRL2 and SMARTS spectral model DNI spectra for identical atmospheric conditions.

1976 U.S. standard atmosphere
Rural aerosol profile
AOD at 500 nm = 0.084 (rural profile)
Total precipitable water vapor = 1.42 atm-cm
Total column ozone = 0.34 atm-cm
Tilt angle of surface (for global hemispherical) = 37.0°
Air mass = 1.5

Note: SPCTRL2 using latitude 40° N, longitude 105° W, day of year 60, time = 11:45 a.m. establishes AM 1.5 (solar zenith angle = 48.2°; 13° incidence angle on tilted surface, assuming surface sloped to south).

Figure 7.2 shows the results of both spectral models for the direct normal spectral irradiance distributions between 300 and 2000 nm. The thick line (SPCTRL2) and thin line (SMARTS) practically overlay each other, with the SMARTS results showing the higher spectral step resolution.

7.9 SPECTRAL DISTRIBUTIONS UNDER ALL SKY CONDITIONS

As for broadband solar radiation modeling for all sky conditions, the impact of clouds on spectral models is quite complex. Complex spectral models such as MODTRAN, libRadtran, and the like address the technical issues and physics but at the cost of requiring extensive detailed information about droplet (or ice crystal) type, sizes, optical properties, or distribution with altitude. Frequently, an assumption is made that clouds operate as spectral "neutral density" (constant attenuation with respect to wavelength) filters, with attenuation proportional to cloud type or layer structure and thickness. However, research has shown that there are significant spectral variations depending on many cloud parameters.

There have been efforts to develop "cloud cover" modification functions for the simple clear sky models such as SPCTRL2 [43]. Derived from data collected by the "Solar Energy Data Unit ['Einheit' in German] System No. 2" of the Centre for Solar Energy and Hydrogen Research, Stuttgart, Germany, the SEDES2 model of Nann and Riordan [44] used the SPCTRL2 model as a basis for an all-sky spectral model based on normalization of spectra by broadband measurements. The CIE also prepared a report developed by Justus and Paris [46] on spectral distributions for various clear sky and cloudy conditions with varying cloud optical depths [45]. As of 2012, the CIE 85 publication is in the process of multilevel technical committee, division, and national membership balloting and comment.

7.10 SUMMARY

The measurement and modeling of the interaction of the AM0 spectral distribution of solar energy with the physical atmosphere are daunting tasks. A great deal of information is needed to accurately model the physics of the interactions between atmospheric constituents and photons originating from the sun. The application of spectral data is important where spectral sensitivities of a renewable energy system are especially great.

The principles of the most complex models were outlined and the details of the CIE daylighting model and Bird's straightforward and simple spectral mode, SPCTRL2, were presented. An overview was provided of the popular moderately complex SMARTS model of Gueymard. Details of the SMARTS models are beyond the scope of this work, but the model is relatively easy to use. SMARTS and SPCTRL2 are available for free over the Internet. These models are relatively accurate for clear sky modeling as long as the input data are reasonable. Given the issues with clear sky spectral modeling, the modeling of terrestrial spectra under all sky conditions is quite a bit more problematic and a subject for further research.

REFERENCES

1. Bird, R.E., R.L. Hulstrom, A.W. Kliman, and H.G. Eldering. (1982). Solar spectral measurements in the terrestrial environment. *Applied Optics*, Vol. 21, No. 8, pp. 1430–1438.
2. Chandreskar, S. (1960). *Radiative Transfer*. Dover, New York.
3. Anon. (1976). *U.S. Standard Atmosphere, 1976.* NOAA/NASA/USAF, Washington, DC. http://www.pdas.com/refs/us76.pdf. Accessed 21 July 2012.
4. Champion, K.S.A., A.E. Cole, and A.J. Kantor. (1985). Standard and reference atmospheres. Jursa, A.S., ed., *Handbook of Geophysics and the Space Environment*, pp. 14.1–14.43. Air Force Geophysics Laboratory. Hanscom Air Force Base, Lincoln, MA.
5. Blättner, W.G., H.G. Horak, D.G. Collins, and M.B. Wells. (1974). Monte Carlo studies of the sky radiation at twilight. *Applied Optics*, Vol. 13, pp. 534–537.
6. Fenn, R.W. et al. (1985). Optical and infrared properties of the atmosphere. Jursa, A.S., ed., *Handbook of Geophysics and the Space Environment*, pp. 18.1–18.80. Air Force Geophysics Laboratory, Hanscom Air Force Base, Lincoln, MA.

7. Scot, N.A., and A. Chedin. (1981). A fast line by line method for atmospheric absorption computations: The automatized absorption atlas. *Journal of Applied Meteorology*, Vol. 20, pp. 802–812.

8. Anderson, G.P., F.X. Kneizys, J.H. Chetwynd, Jr., L.S. Rothman, M.L. Hoke, A. Berk, L.S. Bernstein, P.K. Acharya, H.E. Snell, E. Mlawer, S.A. Clough, J. Wang, S.-C. Lee, H.E. Revercomb, T. Yokota, L.M. Kimball, E.P. Shettle, L.W. Abreu, J.E. and A. Selby. (1996). Reviewing atmospheric radiative transfer modeling: New developments in high and moderate resolution FASCODE/FASE and MODTRAN. Optical Spectroscopic Techniques and Instrumentation for Atmospheric and Space Research II, SPIE 2830. Society of Photo-Optical Instrumentation Engineers, Bellingham, WA.

9. Rothman, L.S., R.R. Gamache, R.H. Tipping, C.P. Rinsland, M.A.H. Smith, D.C. Benner, V.M. Devi, J.-M. Flaud, C. Camy-Peyret, A. Perrin, A. Goldman, S.T. Massie, L.R. Brown, and R.A. Totht. (1992). The HITRAN molecular database: Editions of 1991 and 1992. *Journal of Quantitative Spectroscopy and Radiative Transfer*, Vol. 48, pp. 469–507.

10. Anderson, G.P., J. Wang, M.J. Hoke, F.X. Kneizys, J.H. Chetwynd, L.S. Rothman, L.M. Kimball, R.A. McClatchey, E.P. Shettle, S. Clough, W.O. Gallery, L.W. Abreu, and J.E.A. Selby. (1994). History of one family of atmospheric radiative transfer codes. Passive Infrared Remote Sensing of Clouds and the Atmosphere II, SPIE 2309. 170–183. Society of Photo-optical Instrumentation Engineers, Bellingham, WA.

11. Mayer, B., and A. Kylling. (2011) Technical Note: *The libRadtran Software Package for Radiative Transfer Calculations*. User manual available in PDF format from http://www.libradtran.org/doc/libRadtran.pdf. Accessed 21 July 2012.

12. Drummond, A.J., and M.P. Thekaekara (eds.). (1973). *The Extraterrestrial Solar Spectrum*. Institute of Environmental Sciences, Mount Prospect, IL.

13. Wehrli, C. (1985). *Extraterrestrial Solar Spectrum*. Publication No. 615. Physikalisch-Meteorologisches Observatorium, World Radiation Center (PMO/WRC), Davos Dorf, Switzerland.

14. Neckel, H., and D. Labs. (1981). Improved data of solar spectral irradiance from 0.33 to 1.25 um. *Solar Physics*, Vol. 74, pp. 231–249.

15. Gueymard, C.A. (2006). Reference solar spectra: their evolution, standardization issues, and comparison to recent measurements. *Advances in Space Research*, Vol. 37, pp. 323–340.

16. Harder, J., G.M. Lawrence, G. Rottman, and T. Woods. (2000). Solar spectral irradiance monitor (SIM). *Metrologia*, Vol. 37, pp. 415–418.

17. Harder, J., G. Lawrence, J. Fontenla, G. Rottman, and T. Woods. (2005). The spectral irradiance monitor: Scientific requirements, instrument design, and operation modes. *Solar Physics*, Vol. 230, No. 1, pp. 141–167.

18. Kurucz, R.L. (1995). The solar irradiance by computation. *Proceedings 17th Annual Conference Transmission Models,* Phillips Laboratory, Hanscom AFB, PL-TR-95–2060. G.P. Anderson et al., eds., pp. 333–334.

19. American Society for Testing and Materials. (2006). *Standard Solar Constant and Zero Air Mass Solar Spectral Irradiance Tables*. ASTM E490-00a. ASTM International, West Conshohocken, PA.

20. Woods, T.N., et al. (1996). Validation of the UARS solar ultraviolet irradiances: Comparison with the ATLAS 1 and 2 measurements. *Journal of Geophysical Research*, Vol. 101, No. D6, pp. 9541–9569.

21. Smith, E.V.P., and D.M. Gottlieb. (1974). Solar flux and its variations. *Space Science Review*, Vol. 16, pp. 771–802.

22. American Society for Testing and Materials. (2008). *Test Methods for Electrical Performance of Nonconcentrator Terrestrial Photovoltaic Modules and Arrays Using Reference Cells*. ASTM E 1036-08. ASTM International, West Conshohocken, PA.

23. American Society for Testing and Materials. (2008). *Tables for Reference Solar Spectral Irradiances: Direct Normal and Hemispherical on 37° Tilted Surface.* ASTM G173-06. ASTM International, West Conshohocken, PA.

24. Gueymard, C., D.R. Myers, and K. Emery. (2002). Proposed reference irradiance spectra for solar energy systems testing. *Solar Energy*, Vol. 73, No. 6, pp. 443–467.

25. Gueymard, C. (2001). Parameterized transmittance model for direct beam and circumsolar spectral irradiance. *Solar Energy*, Vol. 71, No. 5, pp. 325–346.

26. Kurtz, S.R., D. Myers, T. Townsend, C. Whitaker, A. Maish, R. Hulstrom, and K. Emery. (2000). Outdoor rating conditions for photovoltaic modules and systems. *Solar Energy Materials and Solar Cells*, Vol. 62, No. 4, pp. 379–391.

27. International Electrotechnical Commission. (2008). *Measurement Principles for Terrestrial Photovoltaic (PV) Solar Devices with Reference Spectral Irradiance Data.* IEC 60904-03. International Electrotechnical Commission, Geneva, Switzerland.

28. American Society for Testing and Materials. (2009). *Standard Test Method for Electrical Performance of Concentrator Terrestrial Photovoltaic Modules and Systems under Natural Sunlight.* ASTM E2527-09. ASTM International, West Conshohoken, PA.

29. Myers, D. (2011). *Review of Consensus Standard Spectra for Flat Plate and Concentrating Photovoltaic Performance.* NREL Report No. TP-5500-51865. National Renewable Energy Laboratory, Golden, CO. 33 pp.

30. International Organization for Standardization and Bureau central de la CIE. (2006). *CIE Standard Illuminants.* Joint ISO/CIE Standard ISO 11664-2:2007/CIE S014/E-2006. International Organization for Standardization, Geneva, Switzerland; Bureau central de la CIE, Paris.

31. International Organization for Standardization and Bureau central de la CIE. (2004). *Standard Method of Assessing the Spectral Quality of Daylight Simulators for Visual Appraisal and Measurement of Colour.* Joint ISO/CIE Standard ISO 23603:2005(E)/CIE S012/E:2004. International Organization for Standardization, Geneva, Switzerland; Bureau central de la CIE, Paris.

32. International Commission on Illumination. (2004). *Colorimetry*, 3rd ed. Technical Report, Publication 15:2004. CIE Central Bureau, Vienna, Austria.

33. Driscoll, W.G., and W. Vaughn (eds.). (1978). *Handbook of Optics*, pp. 9-11 to 9-15. Optical Society of America. McGraw-Hill, New York.

34. American Society for Testing and Materials. (2010). *Standard Specification for Solar Simulation for Terrestrial Photovoltaic Testing.* ASTM E927-10. ASTM International, West Conshohoken, PA.

35. Bird, R.E., and C. Riordan. (1986). Simple solar spectral model for direct and diffuse irradiance on horizontal and tilted planes at the Earth's surface for cloudless atmospheres. *Journal of Climatology and Applied Meteorology*, Vol. 25, pp. 87–97.

36. Bird, R.E., and C. Riordan. (1984). *Simple Solar Spectral Model for Direct and Diffuse Irradiance on Horizontal and Tilted Planes at the Earth's Surface for Cloudless Atmospheres.* SERI Technical Report TR-2436. Solar Energy Research Institute, Golden, CO. 26 pp. http://www.nrel.gov/docs/legosti/old/2436.pdf. Accessed 22 July 2012.

37. Bird, R.E., and R.L. Hulstrom. (1983). Terrestrial solar spectral data sets. *Solar Energy*, Vol. 30, No. 6, pp. 563–573.

38. Garrison, J.D., and G.P. Adler. (1990). Estimation of precipitable water over the United States for application to the division of solar radiation into its direct and diffuse components. *Solar Energy*, Vol. 44, No. 4, pp. 225–241.

39. Jacovides, C.P., D.G. Kaskaoutis, F.S. Tymvios, D.N., and Asimakopoulos. (2004). Application of SPCTRAL2 parametric model in estimating spectral solar irradiances over polluted Athens atmosphere. *Renewable Energy*, Vol. 29, pp. 1109–1119.

40. Kaskaoutisa, D.G., H.D. Kambezidis, A.D. Adamopoulos, and P.A. Kassomenos. (2006). Comparison between experimental data and modeling estimates of aerosol optical depth over Athens, Greece. *Journal of Atmospheric and Solar-Terrestrial Physics*, Vol. 68, pp. 1167–1178.

41. Michalsky, J.J., P.W. Kiedron, Q.-.L. Min, and L.C. Harrison. (2003). Shortwave clear-sky diffuse irradiance in the 300- to 1100-nm range: Comparison of models with UV-VIS-NIR and broadband radiometer measurements at the Southern Great Plains ARM site in September/October 2001. Proceedings of Ultraviolet Ground- and Space-based Measurements, Models, and Effects III, SPIE Conference, 5156, pp. Society of Photo-optical Instrumentation Engineers, Bellingham, WA.

42. Gautier, C., and M. Landsfeld. (1997). Surface solar radiation flux and cloud radiative forcing for the Atmospheric Radiation Measurement (ARM) Southern Great Plains (SGP): A satellite, surface observations, and radiative transfer model study. *Journal of the Atmospheric Sciences*, Vol. 54, No. 10. pp. 1289–1307.

43. Bird, R.E., C.J. Riordan, and D.R. Myers. (1987). *Investigation of a Cloud-Cover Modification to SPCTRAL2, SERI's Simple Model for Cloudless-Sky, Spectral Solar Irradiance*. NREL Report No. TR-215-3038. National Renewable Energy Laboratory, Golden, CO. 37 pp.

44. Nann, S., and C.J. Riordan. (1991). Solar spectral irradiance under clear and cloudy skies: Measurements and a semiempirical model. *Journal of Applied Meteorology*, Vol. 30, pp. 447–462.

45. International Commission on Illumination. (1989). *Solar Spectral Distributions*. Technical Report, CIE Publication No. 85. CIE Central Bureau, Vienna, Austria.

46. Justus, C.G., and M.V. Paris. (1988). A cloudy sky radiative transfer model suitable for calibration of satellite sensors. *Remote Sensing of the Environment*, Vol. 24, pp. 269–285.

8 Introduction to Modeling Daylight

But soft! What light through yonder window breaks?

—**William Shakespeare**, *Romeo and Juliet*, Act 2, Scene 2

8.1 INTRODUCTION

The previous chapters addressed solar radiation applications that use solar irradiance to produce either electricity (photovoltaics) or heat (solar thermal collectors). This chapter addresses the use of natural daylight in place of artificial light for architectural applications. Since it is sometimes desirable to block the direct normal irradiance (DNI) from a window aperture (for heat balance and glare considerations) and the DNI may only occasionally be present in the aperture of a window, architects usually refer to the combination of direct (DNI) and diffuse (diffuse horizontal irradiance, DHI) solar radiation (global hemispherical irradiance, or GHI) solar radiation as "sunlight" and the diffuse solar radiation alone (DHI) as "daylight." The complete evaluation of solar radiation and illumination considerations by architects is a complex task, and beyond the scope of this book. However, a brief overview of daylighting principles of some of the simpler models is provided. An example of a free, rather complete, lighting simulation and rendering program is RADIANCE, described in Reinhart and Andersen [1] and online in detail at http://radsite.lbl.gov/radiance/papers/sg94.1/daylight.html. Other applications, such as the visibility of cell phone screens outdoors or the appearance of materials outdoors, are affected by natural daylight but are beyond the scope of this book.

8.2 ILLUMINANCE VERSUS IRRADIANCE

8.2.1 PHOTOPIC RESPONSE

Previous chapters discussed the modeling of solar components in terms of irradiance, or power per unit area. For applications where natural daylight is desired in place of artificial light, the spectral response of the human eye is taken into account. The International Commission on Illumination, known in French as the *Commission International de l'Eclairage,* or CIE, is the internationally recognized organization devoted to research and standards in the field of illumination. The scientific community recognizes the CIE standard spectral response of the eye for daylight called the *photopic response*. The spectral response of the eye at night (scoptopic response)

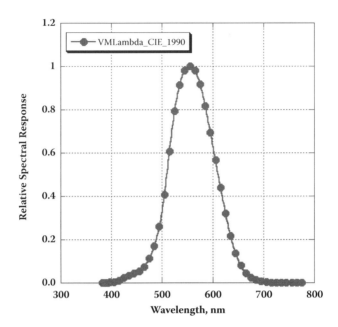

FIGURE 8.1   The photopic response of the human eye as published by the CIE. (Data from SMARTS mode code. Available from http://www.nrel.gov/rredc/smarts.)

is considerably different and is not discussed here. Figure 8.1 shows the CIE human photopic response curve as published in 1990 [2].

Appendix E presents the values of the CIE 1988 modified 2° spectral luminous efficiency function for photopic vision between 380 and 780 nm in 1-nm steps. The table is also available from the CIE on disk in the form of an ASCII file for computers.

The photopic response function is important since it must be multiplied by the spectral source function, and the resulting product function must be integrated over the response function limits to compute the *total luminous flux per unit area* or illuminance of a source. Illuminance is measured in lux (lx) or lumens per square meter. A lumen is equivalent to a candela per steradian (cd sr^{-1}). Since the 16th General Conference on Weights and Measures (CGPM) in 1979, the candela has been defined as follows [3]:

> The candela is the luminous intensity, in a given direction, of a source that emits monochromatic radiation of frequency 540 × 1012 hertz and that has a radiant intensity in that direction of 1/683 watt per steradian. (p. 21. Original in French only.)

The frequency chosen is in the visible spectrum near green, corresponding to a wavelength of about 555 nm. The human eye is most sensitive to this frequency when adapted for bright conditions. At other frequencies, more radiant intensity is required to achieve the same luminous intensity, according to the frequency response of the human eye. The luminous intensity Iv for light of a particular wavelength λ is given by Equation 8.1:

$$Iv(\lambda) = 683 \ V(\lambda) \ I(\lambda) \ lumen \ sr^{-1} = lm \ sr^{-1} \qquad (8.1)$$

where $V(\lambda)$ is the value of the photopic response curve, and $I(\lambda)$ is the source spectral irradiance at wavelength λ (Wsr^{-1}).

Given a source spectral irradiance distribution with units of $Wm^{-2}nm^{-1}$, the illuminance, lumens per square meter, or $lm \ m^{-2}$, Ev, analogous to the irradiance, is given by the integral of Equation 8.1 over all wavelengths:

$$Ev = 683 \int V(\lambda) \ I(\lambda) \ d\lambda \ lm \ m^{-2} \qquad (8.2)$$

Thus, the number of lumens from an incandescent, fluorescent, or light-emitting diode light source as well as natural daylight depends strongly on the spectral distribution of the source. Note that despite the fact that the photopic curve is defined in terms of a specific (2°) solid angle, there is no integration over a solid angle in Equation 8.2 since the definition of the $V(\lambda)$ function photopic curve incorporates the solid angle.

8.2.2 LUMINOUS EFFICACY

The specific wavelength dependence of the definition of the luminous intensity and the comment about the greater radiant intensity needed at different wavelengths for equivalent luminous intensity means that the efficiency of the source in "converting" irradiance to illuminance varies with the source spectral distribution. Converting total irradiance in watts per square meter into illuminance values of lux is not simple because of this dependence on the spectral distribution of the source. The ratio of the total luminous intensity to the total irradiance of a source is the luminous efficacy K of the source:

$$K = Ev/I = (683 \int V(\lambda) \ I(\lambda) \ d\lambda)/(\int I(\lambda) \ d\lambda) \ lm/W \qquad (8.3)$$

The luminous efficacy provides a "conversion factor" to convert more readily available (or modeled) irradiance data into illuminance data. The integrals are performed over all wavelengths.

Examples of the luminous efficacy of various sources as given by McCluney [4] are provided in Table 8.1. It is important to note that the values in Tables 8.2 and 8.3 are representative only and may vary significantly depending on conditions, such as source age or wide variations in atmospheric conditions.

As described in Chapter 7, the spectral distribution of sunlight is itself a strong function of solar elevation or air mass and atmospheric constituents. Approximate values of luminous efficacy for the sun under various conditions are shown in Table 8.2. Table 8.2 shows that there is at least a 40% variation in the luminous efficacy of natural sunlight, depending on solar elevation and the solar component of interest.

The SMARTS (Simple Model of the Atmospheric Radiative Transfer of Sunshine) spectral model mentioned in Chapter 7 includes options for selecting clear sky illuminance and luminous efficacy calculations based on spectra derived from the input

TABLE 8.1
Examples of Luminous Efficacy for Several Lighting Sources

Source	Luminous Efficacy (lm/W)
Incandescent tungsten lamp	15
Tungsten quartz halogen lamp	24
5-W (electrical) light-emitting diode lamp	102
Xenon arc lamp	40
Fluorescent T8 lamp	90
Sun (5800 K; extraterrestrial)	93

Source: Data from Wikipedia http://en.wikipedia.org/wiki/Luminous_efficacy.

TABLE 8.2
Luminous Efficacy for Different Solar Conditions

Solar Condition	Efficacy (lm/W)
Direct sun (low solar elevation)	90
Direct sun (high solar elevation	117
Direct sun (mean solar elevation)	100
Diffuse sky (clear)	150
Diffuse sky (average)	125
Global hemispherical (average)	115

Source: Data from Wikipedia http://en.wikipedia.org/wiki/Luminous_efficacy.

TABLE 8.3
SMARTS Spectral Model Luminous Efficacy for Example Solar Conditions

Air Mass	Aerosol Optical Depth at 500 nm	Pollution	DHI (K lm/W)	GHI (K lm/W)	DNI (K lm/W)
1.5	0.15	Clear	131.5	111.4	108.3
2.5	0.15	Moderate	136.5	108.4	98.7
2.5	0.15	Severe	147.9	106.7	93.6

Source: Data from SMARTS model runs.

data. Representative SMARTS luminous efficacy results for selected atmospheric conditions are shown in Table 8.3. Note the differences between these explicit calculations and the values in Table 8.2. Differences on the order of 15% are apparent.

For these clear sky situations, we see air mass and pollution increase and direct beam and global luminous efficacy decrease; however, the diffuse sky luminous efficacy increases. The situation is not as straightforward under cloudy and partly cloudy conditions as spectral effects of clouds on the spectral irradiance distributions can be more complex.

The use of average values, such as in Table 8.2, or specific calculations using Equation 8.3 in conjunction with spectral models is dependent on the user's desired accuracy for the user's application.

8.3   APPLICATIONS OF DAYLIGHT DATA AND MODELS

8.3.1   INTERIOR APPLICATIONS

In 2010, buildings in the United States consumed 38% of America's energy and 68% of its electricity [5] (see the *Buildings Energy Data Book*, http://buildingsdatabook. eren.doe.gov). The most straightforward application of daylighting is the substitution of daylight for electrical loads for lighting in interior building spaces. This is usually accomplished through window and skylight design and integration into building facades. These applications may also include "light pipe" for collecting and distributing photons from the sky dome to interior spaces [6,7]. Complex daylighting models, such as RADIANCE [1], address daylighting alone. Other total building energy design tools such as Energy-10, developed by the U.S. Department of Energy [8, 9], address total building energy needs as well as the daylighting component. A basic daylighting concept is the *daylight factor*, the ratio of illuminance derived from daylight alone on a work plane of interest (inside) to the illuminance on a similar plane under the hemisphere of a specified (CIE standard) overcast sky. The daylight factor depends on the availability or means of transferring natural daylight to the plane of interest.

Last, there are handbook data for building daylighting applications that provide data in summarized form, such as the National Renewable Energy Laboratory (NREL) *Solar Radiation Data Manual for Buildings* [10]. The NREL data manual presents monthly summaries of incident illuminance for horizontal windows (skylights) and vertical windows facing north, east, south, and west. These values were computed using the Perez anisotropic diffuse model [11] partially described in Chapter 6, Section 6.4, with inputs of global horizontal illuminance, direct beam illuminance, and diffuse horizontal illuminance instead of their solar radiation counterparts. These illuminance data are the input data needed to compute the available daylight values for windows and skylights from the sky dome.

8.3.2   EXTERIOR APPLICATIONS

As briefly mentioned in the introduction to this chapter, exterior applications of illuminance data and models apply mainly to the appearance (to the human eye) of objects outdoors. This may include such applications as the appearance of building facades, display advertisement or informational signs, or menu boards at drive-through establishments. Additional applications include design of automated video and photo image-capturing cameras [12] or the modeling of the readability of a cell phone, tablet computer, and other electronic display screens in varying outdoor conditions. As mentioned in the introduction, there are complex radiance-modeling programs such as RADIANCE [1] and the VELUX Daylight Visualizer software (http://viz.velux.com/daylight_visualizer/download) to address both internal and external applications that are beyond the scope of this book. Exterior applications often depend on the nature of the spatial distribution of sky illuminance over the sky dome. Examples of models addressing the spatial distribution issue are described in Section 8.5. In the next section we describe models for converting available irradiance data into illuminance data.

8.4   THE PEREZ ANISOTROPIC ILLUMINANCE MODEL

A version of the popular model for computing illuminance values uses luminous efficacy models of Perez et al. [11], which were discussed in Chapter 6. Inputs to the luminous efficacy models are global horizontal radiation, direct beam radiation, diffuse horizontal radiation, and dew point temperature. As in Chapter 6, Section 6.4, where the Perez tilt irradiance model was described, sky conditions were partitioned according to a clearness parameter epsilon and sky brightness factor Δ (see Equations 6.13 and 6.14).

$$\varepsilon = [(DHI + DNI)/DHI + 1.041z^3]/[1 + 1.041z^3] \qquad (8.4)$$

where Z is the zenith angle in radians. For Z in degrees, $5.535 \; 10^{-6}$ should be used in place of the 1.041 term.

$$\Delta = m \; DHI/I_0 \qquad (8.5)$$

In this equation, Δ is the sky brightness factor, with m the airmass and DHI and Io the diffuse horizontal and extraterrestrial direct beam irradiances, respectively.

8.4.1   Perez Luminous Efficacy Functions

Global, diffuse, and direct luminance efficacy (G_e, D_e, I_e, respectively) are computed directly from GHI, DHI, and DNI using

$$G_e = GHI[ai + bi \; W + ci \; cos(z) + di \; Ln(\Delta)] \qquad (8.6)$$

$$D_e = DHI[ai, + bi \; W + ci \; cos(z) + di \; Ln(\Delta)] \qquad (8.7)$$

$$I_e = max(0, \; DNI[ai + bi \; W + ci \; e^{(5.73 \; Z - 5)} + di\Delta]) \qquad (8.8)$$

where W is the total precipitable water vapor (atm-cm), and the ai, bi, ci, and di depend on the epsilon bins 1 through 8. These are the same epsilon bins as used in Chapter 6, Section 6.4, for the Perez diffuse model for tilted surfaces (also discussed in the next section). The ai … di coefficients for global, diffuse, and direct luminance efficacy and the illuminance at the zenith are presented in tabular form in Appendix F. Functional fits of the efficacy functions, similar to those of Section 6.4 (Equations 6.15 to 6.20) for

TABLE 8.4
Functional Fits of Appendix F ai ... di Coefficients
for Luminous Efficacy of Solar Components

i = 1.8

Global Luminance Efficacy

Ai	$+0.1338x^5 - 3.0027x^4 + 25.4770x^3 - 100.56x^2 + 175.78x - 1.34$
Bi	$+0.0060x^5 - 0.1463x^4 + 1.3416x^3 - 5.6664x^2 + 10.91x - 6.92$
Ci	$-0.0560x^5 + 1.3253x^4 - 11.9770x^3 + 50.509x^2 - 95.86x + 67.608$
Di	$+0.0723x^5 - 1.6774x^4 + 14.6650x^3 - 59.211x^2 + 105.86x - 68.888$

Diffuse Luminance Efficacy

Ai	$-0.1323x^5 + 2.6945x^4 - 19.092x^3 + 56.594x^2 - 64.726x + 122.29$
Bi	$+0.0027x^5 - 0.0203x^4 - 0.2396x^3 + 2.2606x^2 - 3.5478x + 1.13$
Ci	$+0.0592x^5 - 1.0984x^4 + 7.125x^3 - 19.505x^2 + 12.779x + 12.49$
Di	$-0.0802x^5 + 1.7002x^4 - 12.518x^3 + 37.826x^2 - 45.653x + 10.105$

Direct Luminance Efficacy

Ai	$-0.3728x^4 + 7.8336x^3 - 59.026x^2 + 186.46x - 97.001$
Bi	$-0.0527x^4 + 0.9250x^3 - 5.1477x^2 + 10.482x - 10.661$
Ci	$+0.0068x^5 - 0.1730x^4 + 1.6520x^3 - 7.2701x^2 + 14.359x - 11.548$
Di	$+1.0026x^4 - 19.343x^3 + 131.86x^2 - 378.97x + 382.3$

Zenith Illuminance (lm)

Ai	$2.2944x^2 - 20.342x + 58.725$
Bi	$-0.1781x^3 + 3.3461x^2 - 22.662x + 47.514$
Ci	$-7.5579x^2 + 89.3x - 102.78$
Di	$-1.0904x^3 + 5.0388x^2 + 17.48x - 67.334$

Source: Sandia National Laboratory, data from the development and verification of the Perez diffuse radiation model. Contractor Report Sand88-7030, Sandia National Laboratories, Albuqurque, NM. http://prod.sandia.gov/techlib/access-control.cgi/1988/887030.pdf.

producing the ai ... di coefficients as a function of epsilon bin (1, 2, ... , 8) are given in Table 8.4.

If precipitable water vapor is not available, it can be estimated using the Garrison approximation based on relative humidity (Equations 3.14 to 316 of Section 3.4.1, Chapter 3), or Perez proposed the following approximation based on dew point temperature Td (°C):

$$W = e^{(0.07\,Td - 0.075)} \text{ atm-cm} \tag{8.9}$$

8.4.2 ILLUMINANCE ON TILTED SURFACES

In the same fashion as for the irradiance version of the Perez model, computing the diffuse illuminance on a tilted surface depends on the application of the "anisotropic coefficients" F_1 and F_2 described in Section 6.4, Equation 6.12 and repeated here:

TABLE 8.5
Coefficients for Perez Model Fij Parameters (Compare with Table 6.2)

ε Bin	ε Low	ε High	F11	F12	F13	F21	F22	F23
1	1.000	1.065	0.011	0.570	−0.081	−0.095	0.158	−0.018
2	1.065	1.230	0.429	0.363	−0.307	0.050	0.008	−0.065
3	1.230	1.500	0.809	−0.054	−0.442	0.181	−0.169	−0.092
4	1.500	1.950	1.014	−0.252	−0.531	0.275	−0.350	−0.096
5	1.950	2.800	1.282	−0.420	−0.689	0.380	−0.559	−0.114
6	2.800	4.500	1.426	−0.653	−0.779	0.425	−0.785	−0.097
7	4.500	6.200	1.485	−1.214	−0.784	0.411	−0.629	−0.082
8	6.200	...	1.170	−0.300	−0.615	0.518	−1.892	−0.055

Source: Sandia National Laboratory, data from the development and verification of the Perez diffuse radiation model. Contractor Report Sand88-7030, Sandia National Laboratories, Albuqurque, NM. http://prod.sandia.gov/techlib/access-control.cgi/1988/887030.pdf.

$$DTI = DHI[(1 - F_1) \cos^2(\beta/2) + F_1(a_o/a_1) + F_2 \sin(\beta)] \qquad (8.10)$$

$a_o = \cos \theta_i$, and $a_1 = \cos z$, where θ_i is the DNI incidence angle, Z is the zenith angle, and β is the tilt angle of the surface.

The F_1 and F_2 terms in Equation 8.10 are computed from (note units conversion of the zenith angle in degrees to radians)

$$F_1 = max[0, F11 + F12\Delta + F13z(\pi/180)] \qquad (8.11)$$

$$F_2 = F21 + F22\Delta + F23z(\pi/180) \qquad (8.12)$$

Except in this case, the coefficients Fij used to derive F_1 and F_2 differ from the values in Table 6.2, namely, the values are given in Table 8.5.

Analogous to the Fij functional fits in Section 6.4 (Equations 6.15 to 6.20), the following polynomial regression results for Fij as a function of epsilon bin (= x) from 1 to 8 can be evaluated by multiple linear regression analysis:

$$F11 = 0.0058x^3 + 0.0304x^2 + 0.3156x - 0.3061 \qquad (8.13)$$

$$F12 = 0.0056x^5 - 0.1175x^4 + 0.9082x^3 - 3.1606x^2 + 4.5325x - 1.603 \qquad (8.14)$$

$$F13 = 0.0019x^4 - 0.0303x^3 + 0.1774x^2 - 0.5689x + 0.3369 \qquad (8.15)$$

$$F21 = 0.001x^4 - 0.0166x^3 + 0.0819x^2 - 0.0146x - 0.1424 \qquad (8.16)$$

$$F22 = -0.0043x^5 + 0.0871x^4 - 0.6519x^3 + 2.189x^2 - 3.381x + 1.9282 \qquad (8.17)$$

$$F23 = 0.00007x^4 - 0.00153x^3 + 0.01653x^2 - 0.08510x + 0.05145 \qquad (8.18)$$

Given either modeled or measured GHI, DNI, or DHI irradiance data, applying the appropriate conversion equation (Equation 8.6, 8.7, or 8.8) (using the appropriate coefficients based on clearness bin from either the tabulated or generating functions for the ai … di) provides the luminous efficacy of the component. The resulting efficacy is multiplied by the component irradiance value to produce the illuminance value for the component. For tilted surfaces, the Fij parameters, sky clearness, and brightness parameters are used to generate F_1 (Equation 8.11) and F_2 (Equation 8.12) for use in Equation 8.10 with the zenith and tilt angles to compute the diffuse illuminance on the tilted surface.

8.4.3 Uncertainty of the Perez Anisotropic Illuminance Model

The validation of the illuminance models was accomplished by comparison with measured data from six measurement stations in the northeastern United States. Table 5 in Perez et al. [11] presents detailed site-by-site mean bias and root mean square (RMS) differences between modeled and measured data for the luminance efficacy functions. Note that illuminance measurements are typically accomplished with a photometer, usually a pyranometer with a silicon-based sensor in conjunction with an optical filter that emulates the photopic response function. Besides the contributing factors for uncertainty in these measurements is the accuracy of the filter spectral match to the photopic response [12]. A summarized version of that table is shown in Table 8.6.

The table shows that the basic uncertainty of the model is on the order of ±6.0% to ±10.0% for luminance efficacy over all conditions, and the zenith luminance model is accurate to at best about ±25%. Evaluation of the Perez model as well as six other, some more complex, models, with up to 324 independent coefficients or parameters, and four independent data sets [13, 14], was in agreement with the assessment with the average mean bias error (MBE) of about ±3% in Table 8.6. All seven models

---

TABLE 8.6

Mean Bias and Root Mean Square Error Percentages for Perez Model at Five Sites and Three Sky Conditions

Sky	MBE Range (%)			RMSE Range (%)		
	Global	**Direct**	**Diffuse**	**Global**	**Direct**	**Diffuse**
Overcast	−1.3 to +1.2	n/a	−1.1 to +1.1	3.5 to 5.7	n/a	2.7 to 6.1
Intermediate	−1.3 to +3.8	−4.3 to +1.2	−1.1 to +3.8	2.7 to 3.7	9.8 to 14.0	5.0 to 9.1
Clear	−1.2 to +1.6	−1.3 to +1.3	−1.4 to +3.5	1.9 to 3.3	4.0 to 5.7	10.2 to 17.3
Average (all)	−1.2 to +1.2	−1.6 to +1.5	−1.1 to +2.3	2.9 to 4.3	5.6 to 9.5	5.9 to 10.6

Sky	Overcast		Intermediate		All	
	MBE	**RMSE**	**MBE**	**RMSE**	**MBE**	**RMSE**
Zenith luminance	−9.0 to + 3.0	17.6 to 25.9	−2.3 to 4.9	21.9 to 29.4	0.3 to 3.6	21.5 to 30.6

Source: Sandia National Laboratory, data from the development and verification of the Perez diffuse radiation model. Contractor Report Sand88–7030, Sandia National Laboratories, Albuqurque, NM. http://prod.sandia.gov/techlib/access–control.cgi/1988/887030.pdf.

evaluated had MBEs within this range. However, the RMS errors for all seven models ranged from ±27% to ±62%, averaging about ±40%. Some of the additional RMS error is attributed to differing instrumentation (Sky scanners using photopic filters, rather than photometers) used at some stations to provide the illuminance data used in the validation.

8.5   INTERNATIONAL COMMISSION ON ILLUMINATION MODELS

8.5.1   CIE Standard Sky Models

The CIE has constructed a set of "standard sky conditions" to provide guidance regarding the distribution of sky "brightness" for "energy conscious window design, daylight calculation methods and computer programs as well as for visual comfort and glare evaluations" [15, p. 359], see also [16]. According to Kittler, Perez, and Darula [15], CIE developed a "standard overcast sky" in 1955, and 20 years later established a CIE standard clear sky [17]. In 1997, the analysis of a wide range of measurements from sky luminance scanners and mapping instrumentation by a CIE technical committee resulted in proposed algorithms for computing relative (to the zenith) spatial distributions of luminance from the sky dome for intermediate conditions between the clear sky and overcast sky standard conditions. The proposed algorithms resulted in 15 standard sky conditions corresponding to different combinations of direct solar radiation, circumsolar solar radiation, and diffuse solar radiation relative magnitudes, including both the clear sky and overcast sky standard distributions.

8.5.2   CIE Gradation and Indicatrix Functions

The CIE relative luminance distribution models can easily be implemented in either computer program codes or spreadsheet programs such as Excel®. The model is based on a set of six "gradation" curves, describing the change in luminance as a function of solar altitude or zenith angle and six "indicatrix" functions. Indicatrix functions describe the theoretical distortion of the uniform hemispherical luminance spatial distribution on the sky dome (hemisphere) as scattering centers (pollution, aerosols, clouds, etc.) are introduced into the radiation field. Table 8.7 describes the 15 standard distributions in qualitative terms and in terms of the gradation and indicatrix function appropriate for each standard sky condition.

The gradation functions are of the form

$$\phi(z)/\phi(0) = [1 + a \exp(b/\cos(z)]/(1 + a \exp(b)) \tag{8.19}$$

where $\phi(z)$ and $\phi(0)$ are the zenith = 0 and zenith = z values determined from zenith angle z and the parameters a and b, and exp is the exponential function $\exp(x) = e^x$. Thus, in Table 8.7, $\phi(0)$ is computed using z = 0 and the values of a and b listed in Table 8.8.

The indicatrix functions represent varying obscuration and scattering and thus a varying circumsolar radiation component in the sky luminance distribution. These functions are of the form

TABLE 8.7
CIE Standard Sky Luminance Distributions

Type	Gradation Function φ()	CIE Definition	Indicatrix Function f()	F(0)
1	I	CIE standard overcast sky	1	2.986
2	I	Overcast steep luminance gradation; slight solar brightening	2	2.986
3	II	Overcast, moderately graded; azimuth uniform	1	1.494
4	II	Overcast, moderately graded; slight solar brightening	2	1.494
5	III	Uniform Illuminance; isotropic sky	1	1.000
6	III	Partly cloudy, no gradation to zenith; slight solar brightening	2	1.000
7	III	Partly cloudy, no gradation to zenith; brighter circumsolar solar brightening	3	1.000
8	III	Partly cloudy, no gradation to zenith; distinct solar corona	4	1.000
9	IV	Partly cloudy, obscured sun	2	0.423
10	IV	Partly cloudy, bright sun	3	0.423
11	IV	White-blue with distinct corona	4	0.423
12	V	CIE standard clear sky with low luminance and high turbidity	4	0.274
13	V	CIE standard clear sky with pollution	5	0.274
14	VI	Cloudy less-turbid sky with broad corona	5	0.139
15	VI	White-blue turbid sky with broad corona	6	0.139

Source: Lawrence Livermore Laboratories. Data adapted from http://radsite.lbl.gov/radiance/man_html/ gensky.1.html.

TABLE 8.8
Gradation and Indicatrix Function Parameters

Brightness Gradation	a	b	Indicatrix Type	c	d	e
I	4.0	−0.70	1	0	−1.0	0.00
II	1.1	−0.80	2	2	−1.5	0.15
III	0.0	−1.00	3	5	−2.5	0.30
IV	0.0	−0.55	4	10	−3.0	0.45
V	−1.0	−0.32	5	16	−3.0	0.30
VI	−1.0	−0.15	6	24	−2.8	0.15

Source: Adapted from http://mathinfo.ens.univerims.fr/IMG/pdf/other2.pdf.

TABLE 8.9

Complete Parameters for Each of Fifteen CIE Standard Sky Luminance Distributions

Sky Type (Table 8.7)	Gradation Function	a	B	Indicatrix Function	c	d	e
1	I	4.0	−0.70	1	0	−1.0	0.00
2	I	4.0	−0.70	2	2	−1.5	0.15
3	II	1.1	−0.80	1	0	−1.0	0.00
4	II	1.1	−0.80	2	2	−1.5	0.15
5	III	0.0	−1.00	1	0	−1.0	0.00
6	III	0.0	−1.00	2	2	−1.5	0.15
7	III	0.0	−1.00	3	5	−2.5	0.30
8	III	0.0	−1.00	4	10	−3.0	0.45
9	IV	−1.0	−0.55	2	2	−1.5	0.15
10	IV	−1.0	−0.55	3	5	−2.5	0.30
11	IV	−1.0	−0.55	4	10	−3.0	0.45
12	V	−1.0	−0.32	4	10	−3.0	0.45
13	V	−1.0	−0.32	5	16	−3.0	0.30
14	VI	−1.0	−0.15	5	16	−3.0	0.30
15	VI	−1.0	−0.15	6	24	−2.8	0.15

Source: Adapted from http://mathinfo.ens.univerims.fr/IMG/pdf/other2.pdf.

$$f(\chi) = 1 + c[\exp(d\chi) - \exp(d\ \pi/2)] + e\ \cos^2(\chi) \tag{8.20}$$

The variable χ is the angular distance between the position of the sun (zenith and azimuth) and the location of a patch of sky at a different azimuth and zenith location. The parameters c, d, and e for types 1 to 6 are listed in Table 8.8 along with the coefficients for the gradation functions.

Table 8.9 is a complete summary of the 15 sky distributions and the appropriate coefficients for use in Equations 8.19 and 8.20.

The standard formula for computing the relative luminance distribution for any standard sky is

$$L = Lz\ [f(\chi)\ \phi(z)]/[f(Zs)\ \phi(0°)] \tag{8.21}$$

L is the luminance of the sky element at angular distance χ from the sun. See Figure 8.2 for the geometry; $f(\chi)$ and $f(Zs)$ are the indicatrix function (Equation 8.20) for the sky type at solar zenith position Zs. $\phi(0)$ and $\phi(z)$ are the gradation functions (Equation 8.19) at the zenith = 0 and z = sky patch zenith angle.

By dividing up the sky in a matrix of azimuth and zenith angle steps and computing the solar azimuth and altitude, one may compute the angular distance χ between the sun and a zenith/azimuth patch of the sky and compute the relative luminance

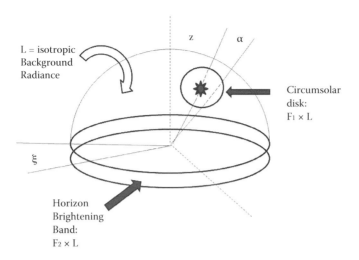

FIGURE 8.2   Geometry for CIE luminance distribution calculations.

value with respect to the zenith. Note that the CIE model does not specify an absolute magnitude or manner of estimating the value of Lz, or $\phi(0°)$.

The solar azimuth and zenith angle of the sun are calculated, and the gradation function $\phi(0)$ is computed. Next, for each patch of sky, say in steps from zenith angle $0°$ to $90°$ and steps of azimuth from $0°$ to $360°$, at the desired angular resolution, the angular separation gradation function $\phi(z)$ and the indicatrix functions $f(\chi)$ and $f(Zs)$ are calculated for each desired location in the sky. The indicatrix functions depend on the angular distance χ between the sky patch of interest and the sun. This is computed using the spherical law of cosines:

$$\cos(\chi) = \cos(Zs) \cos(Zp) + \sin(Zs) \sin(Zp) \cos(|Azp\text{-}Azs|) \qquad (8.22)$$

χ is the angular distance between the sun, at azimuth angle Azs and zenith angle Zp, and the patch of sky at azimuth Azp and zenith angle Zp.

For each computed z and χ, Equation 8.19 is computed using the a and b in Table 8.10, and Equation 8.20 is computed using the coefficients c, d, and e in Table 8.10. The luminance at the zenith Lz is computed from Equation 8.19 for z = 0 and the a and b for the sky type selected.

Finally, the luminance at the patch of sky at Zp, Azp, is computed from Equation 8.21.

8.5.3   COMPUTATIONAL EXAMPLE

We compute an example for 8 a.m. standard time for May 30 at a location of $30°$ N, $105°$ W, at an elevation of 100 m above sea level. Assume the distribution desired is for a white-blue sky with distinct corona. From Table 8.7, the model type selected is 11; the gradation function is type IV, and the indicatrix function is type 4. Table 8.10 lists parameters for the computation, including the day angle for the day of the year (DANG) and zenith and azimuth angles of the sun.

TABLE 8.10

Parameters for Example Sky Patch Luminance Computation

Model No.	Gradation Function	a	b	Indicatrix Function	c	d	e	Model Description
11	IV	−1.0	−0.55	4	10	−3	0.45	White–blue with broad corona

Lat	Lon	Time Zone	Elevation (m)	Station Press (mB)	Annual O_3 (atm–cm)
30	−105	7	100	1013.37	0.3

Month	Day of Month	Hour	DOY	DANG	DEC	EQT	Sun Azs	Zs
5	30	8:00	150	2.565°	18.483°	2.849	115.2°	59.4°

Note: DOY = day of year; DEC = declination, EQT = equation of time.

For the selected location, date, and time, the solar zenith Zs is 59.4° (elevation angle = 30.6°), and azimuth Azs is 115.2° (from north = 0°). Thus, for a nearby patch of sky at Zp = 62° (elevation = 28.0°) and azimuth Azp = 110°, Equation 8.22 gives

$$\cos(\chi) = \cos(59.4°) \cos(62.0°) + \sin(59.4°) \sin(62.0°) \cos(|110.0° - 115.2°|) = 0.9958$$

So, $\chi = 5.21°$ or 0.0910 radians.

Lz, the luminance at the zenith, from Equation 8.19 with z = 0, $\phi(0) = 1.0 - e^{-0.55}$ = 0.423.

The indicatrix function at the solar zenith angle f(Zs), where Zs = 59.4° is computed from Equation 8.20 with $\chi = 59.4° = 1.037$ radians using c = 10.0, d = −3.0, and e = 0.45:

$$f(Zs = 59.4° = 1.037 \text{ rad}) = 1 + 10.0[e^{-3.11} - e^{-4.712}] + 0.45 (0.50904)^2 = 1.557$$

For the sky patch at zenith angle 62° = 1.082 radians, the gradation value (Equation 8.19) using a = −1.0, b = −0.55 is

$$\phi(Zp = 62.0° = 1.082 \text{ rad}) = 0.423[1 - e^{-0.55/0.469}]/(1 - e^{-0.55}) = 0.690$$

and the indicatrix value at the sky patch 5.21° from the solar center (Equation 8.20) with $\chi = 5.21°$ is

$$f(\chi = 5.21° = 0.0910 \text{ rad}) = 1 + 10.0[e^{-2.909} - e^{-7.068}] + 0.45(0.9959)^2 = 8.967$$

and from Equation 8.21

$$L(62°, 110°) = 0.423[(8.967)(0.690)/(1.557)(0.423)] = 3.974$$

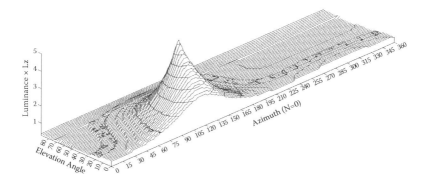

FIGURE 8.3   Topographic plot of luminance distribution as function of elevation (90° zenith angle) and azimuth Az for CIE sky type 11 at 30 N, 105 W, 8:00 a.m. on May 30.

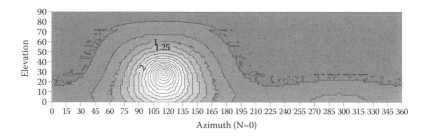

FIGURE 8.4   Contour plot of luminance distribution as function of elevation angle and azimuth angle for same conditions as Figure 8.2. Contours are relative magnitude with respect to Lz.

Topographic and contour plots in elevation-azimuth coordinates for the entire sky hemisphere for the conditions in Table 8.10 are shown in Figures 8.3 and 8.4.

8.6   SKY LUMINANCE MODEL ACCURACY

Lam, Mahdavi, and Pal [18] discussed the accuracy of the CIE and Perez models and three others. They used measured diffuse sky irradiance data to obtain the estimated magnitude of the luminance at the zenith. Their evaluation compared illuminance levels measured inside an actual daylight space and on the roof of the building. The basis of the comparison was the daylight factor, described in Section 8.2.1, derived from the data and the models. These authors found all six of the models had individual average systematic errors of ±20%. The average systematic bias of all six models was –10%, and all six models had similar random error dispersion of an additional ±20%.

The model with the smallest systematic error was the Perez model, with about –2.5% systematic error. The CIE models (overcast and those based on the Kittler et al. approach) showed about –5% to –10% systematic differences with respect to the measured data. Another analysis performed by Mardaljevic [19] resulted in very similar results. It is a good idea to remember that the absolute errors related to these relatively large percentage errors are rather reasonable. For instance, a 10% error in a typical diffuse sky irradiance with magnitude of 300 Wm^{-2} is only 30 Wm^{-2}, only about three times the uncertainty in the measurements of diffuse irradiance described in Chapter 2, Section 2.6.5.

8.7   OTHER SKY LUMINANCE DISTRIBUTION MODELS

Many arbitrary sky conditions other than the 15 CIE "standard" conditions occur in nature, so attempts have been made to develop models relating general sky conditions to the CIE standard conditions. Examples were developed by Igawa et al. [20] and Darula and Kittler [21]. The approach is generally to relate the ratio of zenith luminance to the global hemispherical or diffuse hemispherical irradiance (GHI, DHI) to establish the zenith illuminance. The issue then becomes which sky type to select and how to modify the sky type to produce the desired general sky luminance distribution. In accordance with Igawa's approach, the following steps are needed: The authors slightly modified the relation for the CIE standard sky calculation gradation and indicatrix functions (Equations 8.19 and 8.20). They rewrote the gradation function as a function of elevation angle e_s with only one parameter, a:

$$\Phi(es) = 1 + a(1.0 - \sin(e_s)^{0.6}) \tag{8.23}$$

and the indicatrix functions as a function of angle (solar elevation e_s or χ between sun and sky patch) using only three parameters, b, c, and d:

$$\psi(\chi) = 1.0 + b[\exp(c\chi) - \exp(c\,\pi/2)] + d\,\cos(\chi)^2 \tag{8.24}$$

So, the luminance as a function of solar elevation angle es, angle between sun and sky patch χ, elevation of the sky patch e_p, and luminance at the zenith becomes

$$L(e_s, e_p, \chi) = Lz[\Phi(e_p)\,\psi(\chi)]/[\Phi(\pi/2)\,\psi(\pi/2 - e_s)] \tag{8.25}$$

A standard relative global illuminance Lgs is computed as a function of solar elevation angle e_s:

$$Lgs = -0.323\,e_s^4 + 1.486\,e_s^3 - 2.581\,e_s^2 + 2.090\,e_s + 0.190 \tag{8.26}$$

Next, a normalized global (hemispherical) illuminance Ngl is computed based on the relative global illuminance G_L and air mass m, where

$$G_L = m\,GHL/GoL \tag{8.27}$$

where GHL is the global hemispherical illuminance, and GoL is the extraterrestrial (direct normal) illuminance (from Table 8.1, GoL is nominally (93 lm/w)*1366 Wm^{-2} ~ 127.038 kLx). Note that GHL can be estimated from irradiance data using Perez's model for global luminous efficacy given in Table 8.4 for the appropriate sky clearness bin. Of course, the zenith illuminance Lz may also be computed from Perez's model (last entry in Table 8.4).

The normalized global illuminance is then

$$Ngl = G_l/Lgs \qquad (8.28)$$

The parameter Ngl is the basis of the Igawa et al. model, in that the coefficients a, b, c, d, and e used in the CIE standard sky models can be computed as functions of Ngl. The result of 300 regression analysis fits of the coefficients against Ngl were (let x = Ngl in all equations) as follows:

$$a = 9.93x^3 - 10.68x^2 + 7.09x - 2.11 \qquad (8.29)$$

$$b = 23.4 (1.6x)^{5.9} \exp(-0.17x) (1.1 - x)^{1.5} \qquad (8.30)$$

$$c = 62.16x^6 - 257.62x^5 + 405.67x^4 - 296.6x^3 + 99.3x^2 - 16.34x + 0.43 \qquad (8.31)$$

$$d = 2.06x^5 - 6.04x^4 + 6.02x^3 - 1.31x^2 + 0.09x \qquad (8.32)$$

Igawa et al. also developed an independent model for calculating the absolute value of the zenith illuminance using the Ngl parameter and the solar elevation angle e (again using x = Ngl)

$$Lz = \exp(Ax^5 + Bx^4 + Cx^3 + Dx^2 + Ex + F) \, lx \qquad (8.33)$$

where the coefficients A, B, C, D, E, and F are functions of the solar elevation angle e_s:

$$A = 18.373e_s + 9.955 \qquad (8.34)$$

$$B = -52.013e_s - 37.766 \qquad (8.35)$$

$$C = 46.457e_s + 59.352 \qquad (8.36)$$

$$D = 1.691e_s^2 + -16.498e_s - 48.670 \qquad (8.37)$$

$$E = 1.124e_s + 19.738 \qquad (8.38)$$

$$F = 1.170 \ln(e_s) + 6.369 \qquad (8.39)$$

The resulting "all-sky model" uses these equations for Lz and the coefficients a, b, c, d, and e for the indicatrix and gradation functions in the modified standard CIE distribution Equation 8.26.

8.7.1 COMPUTATIONAL EXAMPLE

In the numerical calculation for standard sky type 11 used in Section 8.5.3, the solar elevation angle $90 - 59.4 = 30.6°$ (0.534 radians). The air mass m for this geometry is m = 1.9645. Assume the global irradiance GHI is 425 Wm^{-2}. Using average luminous efficacy of GHI from Table 8.2 (115 lm/W), GHL = 48.875 kLx. From Equation 8.26,

$$Lgs = -0.323(0.534)^4 + 1.486(0.534)^3 - 2.581\ (0.534)^2 + 2.090(0.534) + 0.190 = 0.770$$

from Equation 8.27,

$$G_L = m\ GHL/GoL = (1.964*71.875)/(127.038) = 0.756$$

and from Equation 8.28,

$$Ngl = 0.756/0.770 = 0.981$$

From Equations 8.31 to 8.36, A = 19.767, B = –65.54, C = 84.16, D = –56.999, E = 20.33, and F = 5.635, so

$$Lz = 1710\ lx\ or\ 1.710\ klx.$$

To compute the luminance at the same sky patch (c = 5.21° = 0.0910 radians removed from the sun) used in the example in Section 8.4.3 using these assumptions,

From Equations 8.29 to 8.32, a = 3.948; b = 11.590; c = –3.000; d = 0.796.

From Equations 8.23 and 8.24, for ep = 90° – 62° = 28° = 0.4887 radians, and c = 5.21° = 0.0910 radians:

$$\Phi(e_p) = 1 + 3.948(1.0 - \sin(0.4887e_p)^{0.6}) = 2.556$$

$$\psi(\chi) = 1.0 + 11.59[\exp(-3.0(0.0910)) - \exp(-3.0(\pi/2)] + 0.796\ \cos(0.0910)^2 = 10.509$$

$$\Phi(\pi/2) = 1.0$$

and

$$\pi/2 - e_s = 90° - 59.4° = 30.6° = 0.534\ rad$$

$$\psi(\chi) = 1.0 + 11.59[\exp(-3.0(0.534)) - \exp(-3.0(\pi/2)] + 0.796\ \cos(0.534)^2 = 6.206$$

Then, from Equation 8.25, using Lz = 1.710 klx,

$$L(e_s, e_p, \chi) = 1.710[(2.556)(10.506)]/(6.206)] = 7.405\ klx.$$

The relative luminance computed from the "pure" type 11 CIE standard sky of L(patch) = 3.97 in Section 8.4.3 multiplied by the Lz of 1.710 klx estimated here produces a luminance at the patch of 6.789 lx, or a difference of about 8%.

8.8 OTHER READING

The International Building Performance Simulation Association is a good resource for past and present efforts regarding daylighting (and building simulation in general). Its Web site (http://www.ibpsa.org/m_papers.asp, accessed 10 July 2012) provides access to extensive conference proceedings papers addressing daylight modeling and applications to building simulations.

8.9 SUMMARY

Given the large, inefficient loads that artificial lighting involves, daylighting strategies can have a significant impact on energy usage. The nonuniform response of the human eye, widely varying solar spectral distribution, and complex spatial distribution of sunlight make the modeling of daylight in the outdoors and especially in interior spaces quite complex. Yet, many of the concepts developed here are applied regularly in producing computer-generated scenes that are almost photographic in their realism. The selection of models reported provide the basic input to many interior lighting models for architectural tools, such as the RADIANCE model. Used in conjunction with the optical properties of construction or appliance materials, these models provide basic input to models for computing the appearance of buildings, appliances, and natural materials in the outdoors, in fact, outdoor scenes in general. While not as straightforward an application of renewable energy concepts as the direct conversion of solar energy into thermal or electrical energy, utilizing daylight to supplant artificial lighting can greatly enhance energy conservation.

REFERENCES

1. Reinhart, C.F., and M. Andersen. (2006). Development and validation of a RADIANCE model for a translucent panel. *Energy and Buildings*, Vol. 38, No. 7, pp. 890-904.
2. International Commission on Illumination. (1990). *CIE 1988 2° Spectral Luminous Efficiency Function for Photopic Vision*. Central Bureau of the CIE, Vienna, Austria.
3. Giacomo, P. (1980). News from the BIPM. *Metrologia*, Vol. 16, No. 1, 55–61. See also http://www.bipm.org/en/CGPM/db/16/3/. Accessed 9 July 2012.
4. McCluney, R. (1994). *Introduction to Radiometry and Photometry*. Artech House, Norwood, MA.
5. D&R International, Ltd. (2012). *Buildings Energy Data Book*. Buildings Technologies Program Energy Efficiency and Renewable Energy, U.S. Department of Energy. Washington, DC.
6. Farrell, A.J., B. Norton, and D. Kennedy. (2004). Lightpipe daylight simulation modeling using Radiance backward and forward ray tracing methods: A comparison with monitored data for commercial lightpipes in Ireland. Third International Radiance Workshop Ecole D'ingenieurs et d'architectes de Fribourg 11–12 October 2004, Fribourg, Switzerland. http://www.radiance-online.org/radiance-workshop3. Accessed 9 July 2012.
7. Abdul-Rahman, H., and C. Wang. (2010). Limitations in current day lighting related solar concentration devices: A critical review. *International Journal of the Physical Sciences*, Vol. 5, No. 18, pp. 2730–2756. http://www.academicjournals.org/IJPS. Accessed 21 July 2012.

8. Balcomb, J.D. (1994). Energy-10: A Design Tool for Smaller Commercial Buildings. Burley, S.M., et al., eds. Passive '94: *Proceedings of the 19th National Passive Solar Conference,* 25–30 June 1994, San Jose, California. Volume 19. American Solar Energy Society, Boulder, CO, pp. 151–156.

9. Walker, A., D. Balcomb, G. Kiss, N. Weaver, and M. Humphry-Becker. (2003). Analyzing two federal building-integrated photovoltaics projects using ENERGY-10 simulations. *Journal of Solar Energy Engineering: Transactions of the American Society of Mechanical Engineers,* Vol. 125, pp. 28–33; NREL Report No. JA-710-34480 Preprint available at http://www.nrel.gov/docs/fy02osti/31310.pdf. Accessed 21 July 2012.

10. Marion, W., and S. Wilcox. (1995). *Solar Radiation Data Manual for Buildings.* NREL/ TP-463-7904. National Renewable Energy Laboratory, Golden, CO. http://www.nrel. gov/docs/legosti/old/7904.pdf. Accessed 9 July 2012.

11. Perez, R., P. Ineichen, R. Seals, J. Michalsky, and R. Stewart. (1990). Modeling daylight availability and irradiance components from direct and global irradiance. *Solar Energy,* Vol. 44, No, 5, pp. 271–289.

12. Alves, J., and C. Gueymard. (2009). Optical engineering application of modeled photo-synthetically active radiation (PAR) for high-speed digital camera dynamic range optimization. Tsai, B.K., ed. *Proceedings Vol. 7410 Optical Modeling and Measurements for Solar Energy Systems III.* Society for Photo-optical Instrumentation Engineers, (SPIE), Bellingham, WA.

13. Ineichen, P., B. Molineaux, and R. Perez. (1994). Sky luminance data validation: Comparison of seven models with four data banks. *Solar Energy,* Vol. 52. No. 4, pp. 337–346.

14. Olseth, J.A., and A. Skartveit. (1989). Observed and modeled luminous efficacies under arbitrary cloudiness. *Solar Energy,* Vol. 42, pp. 221–233.

15. Kittler, R., R. Perez, and S. Darula. (1997). A new generation of sky standards. *Proceedings Conference Lux Europa,* pp. 359–373.

16. International Organization for Standardization and International Commission on Illumination. (2004). *Spatial Distribution of Daylight—CIE Standard General Sky.* Joint ISO/CIE Standard ISO 15469:2004 (E)/CIE S 011/E:2003. CIE Central Bureau, Vienna, Austria.

17. Commission Internationale de l'Eclairage. (1973). *Standardisation of Luminance Distribution on Clear Skies.* CIE Publication No. 22. Commission Internationale de l'Eclairage, Paris.

18. Lam, K.P., A. Mahdavi, and V. Pal. (1997). The implications of sky model selection for the prediction of daylight distribution in architectural spaces. *Proceedings of International Building Performance Simulation Association,* 8–10 September, Prague, Czech Republic. http://www.ibpsa.org/proceedings/BS1997/BS97_P042.pdf. Accessed 21 July 2012.

19. Mardaljevic, J. (1999). Daylight Simulation: Validation, Sky Models and Daylight Coefficients, PhD. thesis, Institute of Energy and Sustainable Development De Montfort University, Leicester. http://climate-based-daylighting.com/doku.php?id= resources:thesis. Accessed 9 July 2012.

20. Igawa, N., H. Nakamura, and K. Matsuura. (1999). Sky luminance distribution model for simulation of daylit environment. *Proceedings of International Building Performance Simulation Association,* 13–15 September, Kyoto, Japan. http://www.ibpsa.org/proceedings/ BS1999/BS99_PB-01.pdf. Accessed 9 July 2012.

21. Darula, S., and R. Kittler. (1999). *CIE General Sky Standard Defining Luminance Distributions.* http://mathinfo.ens.univ-reims.fr/IMG/pdf/other2.pdf. Accessed 21 July 2012.

9 Summary and Future Prospects

If you don't know anything about computers, just remember that they are machines that do exactly what you tell them but often surprise you in the result.

—Richard Dawkins, 1996

9.1 OVERVIEW OF THE MODELING CHAPTERS

The models described in the previous chapters were selected because of the frequency of their appearance in the literature (read, "popularity"), ease of use and implementation, and the frequency of questions about their accuracy. All utilize to some extent a subset of the fundamentals of solar radiation described in Chapter 1. These fundamentals include the concepts of solar position, solar time, air mass, solar components, zenith and incidence angles, the atmospheric filter, and clearness index. All of the models have been developed and validated, most many times over, by comparison with measured solar radiation data, often of unknown quality.

Chapter 2 discussed the equipment and methods of measurements for solar radiation components and the issues of data quality. Repeated here is the comment that no model can produce solar data more accurate than the solar measurements used to develop or validate the model. Thus, knowledge of quality, accuracy, or uncertainty of measured solar data is critical for evaluating model performance. Most of the models have been validated against data sets from a variety of locations and time periods. However, the performance metrics are often disparate. One author may report absolute errors in watts per square meter, another in joules per square meter, another in percentages. Mean bias error (systematic differences) and root mean square errors (random differences) between measured and modeled data are the most often reported performance metrics. These parameters represent "average offsets" and "average fluctuations" from and about measured data. It may be the case that the offsets themselves have a distinct distribution. These performance metrics may be site dependent or a function of the climate of the sites where measured data were obtained. I recommend that careful thought be given to the evaluation of model performance, whether performed as part of a project or extracted from the existing literature. The conclusion regarding measurement accuracy for thermal, broad-spectrum detectors is that most direct normal irradiance (DNI) measurements are accurate to ±2% to ±3% at best. Total hemispherical ("global") irradiance on a horizontal surface (global horizontal irradiance, GHI) may be accurate to ±3% to ±5% for small ranges of zenith or incidence angles but can be very large (>10%) for large

(>75°) incidence angles because of poor cosine response. The same is true for diffuse hemispherical irradiance (diffuse horizontal irradiance, DHI), except that the "percentages" are typically "full scale" (= 1000 Wm^{-2}), so percentage errors in DHI for absolute values of 100 to 500 Wm^{-2} can be considerably larger, especially if the problem of infrared thermal offset is not considered. Nighttime values of irradiance that depart greatly (more than 2 to 3 Wm^{-2}) from zero can be an indication of this problem. Silicon photodiode and other solid-state detectors of the photovoltaic type are typically larger because of their inherent limited spectral response range. The widely varying solar spectral distribution causes similar variations in the accuracy of these detectors. In all events, it is important to remember that even if radiometer responses are characterized and empirically fit and corrected with regression functions, these fits themselves have a distribution of differences, or "errors," about the regression lines. These residuals are themselves usually randomly distributed, with some standard deviation (the standard error of the estimate). Thus, performing corrections may actually increase the uncertainty in a corrected data point.

Chapters 3 and 4 addressed the modeling of solar irradiance under clear sky irradiance (transmittance) and general sky conditions, respectively. The concept of a linear combination (products) of atmospheric constituents was introduced. The most popular empirically derived transmittance functions were outlined. These models are all sensitive to the number and accuracy of input parameters. Depending on the accuracy desired, long-term climatic values of input variables or inputs for each individual desired data point may be needed. Knowledge of the stability and type of climate (low or high variability in parameters such as precipitation, temperature, cloudiness) for the site of interest is crucial in understanding and estimating uncertainty in model results. Sensitivity studies with respect to each parameter known to vary a great deal will improve the user's understanding of the model results. While time consuming, the application of several models for the same site and as many common input parameters as possible is recommended. The range of model outputs obtained can be very illuminating—or confusing.

If one is lucky to have at least some measured data, usually GHI, for a site, the methods of Chapter 5 may be used to estimate the values of missing solar components. However, given that the parameter space for the balancing of the component equation GHI = DNI cos(Z) + DHI is very broad, the estimated result can be highly uncertain. In particular, the lack of knowledge about the position of clouds with respect to the solar disk and DNI beam adds a great deal of random error to the estimates of DNI from GHI. The correlation scatter plot of DHI/GHI with respect to GHI is even greater and highly site/climate dependent. Again, the application of several models to compare results and obtain a range of results can be helpful in characterizing site "solar climate" variability.

Since it is almost impossible and extremely rare to have the luxury of solar irradiance data exactly matching a solar collector configuration for anything other than a horizontal plane or DNI (for concentrating collectors), it is essential to be able to model the irradiance (or illuminance, for daylighting applications); one needs to be able to model the irradiance on an arbitrary tilted surface. Chapter 6 describes the various "tilt conversion models" developed over the past 30 or more years and discusses their relative merits. Models that assume an isotropic diffuse radiation field are rather simple but usually underestimate the diffuse irradiance on a tilted surface. There are several approaches to describing an anisotropic diffuse radiation field, but the Perez tilt conversion model, with horizon brightening and circumsolar algorithms embedded within it, is by far the most often used. This model is often incorporated into both photovoltaic and daylighting performance and design software packages. Almost every independent evaluation of the Perez tilt radiation model or system performance model using it places it in the top two "best performers" in multiple model performance studies. However, the accuracy of the model is still two to three times worse than measured data, up to 15% to 20% or more for various problem scenarios, such as north-facing surfaces. In addition, for any tilt conversion model, good knowledge of ground albedo values, especially for climates with snowy seasons, is essential to obtain ground-reflected diffuse components of any accuracy.

Chapter 7 described the modeling of the distribution of solar power with wavelength or solar spectral power distributions. These models are critical in evaluating the performance, and performance variations, in photovoltaic conversion technologies that have finite and sometimes limited wavelength regimes for their spectral response. The widely varying spectral distribution parameter space results in the need to establish reference spectral distributions. These reference spectra are needed to compare the performance of various technologies or establish if improvements in manufacturing processes improve or degrade photovoltaic cell or module performance with respect to a single spectrum. Changing solar spectra in the natural environment also has an impact on the performance of solid-state photoelectric sensors used in radiometry, increasing their uncertainty when compared with broadband thermal detectors, as described in Chapter 2. As discussed in Chapter 3, the spectral distribution of sunlight, as well as artificial source for lighting, also have an impact on human visual response.

While not a direct energy conversion technology, but rather an energy conservation approach, daylighting can be considered a valuable renewable energy strategy. Concerned almost exclusively with the large extent and relative spatial distribution with respect to the zenith luminance of the diffuse radiation over the sky dome, daylighting computations of daylight factor (ratio of interior illuminance to available exterior illuminance) require an estimate of how the relative spatial distribution varies through time and by sky "type." Thus, the International Commission on Illumination (CIE) has developed "standard sky" distributions and means to map relative spatial distributions as a function of solar zenith angles and 15 sky types. The distribution results from a combination of sky "gradation" and "indicatrix" functions, each computed independently of the other. These relative spatial distributions represent clear, overcast, and "modified" clear sky conditions. To model intermediate sky conditions, several approaches using the standard sky conditions in conjunction with modifiers based on measured (or modeled) diffuse-to-total irradiance ratios. Typically, the accuracy of these models suffers from the same lack of cloud position information as the anisotropic tilt conversion and all-sky condition irradiance models. Thus, the uncertainty in these models can be relatively large (up to ±25% or so).

9.2   CURRENT ISSUES AND FUTURE PROSPECTS

The primary issue for the user of any model is whether it is appropriate for the application or problem under investigation. Does the user have some measured data to evaluate and compare with a model result? Does the user need to evaluate one or more particular collector configurations or designs (Concentrator or flat-plate collector? Optimized or constrained tilt or azimuth aspect for a collector?)? Is a clear sky "maximal envelope" of resources sufficient? Is a realistic time series of data needed? Are the actual input parameters required for a model available? If not, is a suitable surrogate for missing input parameters available? As mentioned many times here and in the individual chapters, the overriding concerns regarding all of the models discussed (and models in general) are accuracy and uncertainty. All of these questions lead us to consider the following needs for future research:

1. Availability and quality of measured solar radiation data must be improved. This includes (a) better radiometer design for more accurate data; (b) more widely distributed, geographically complete measurement stations; (c) higher time resolution data as often 1-s data are needed for studying transient behavior of very large photovoltaic arrays; and (d) more public, easily accessible data (Web portals, data applications).
2. More and better model input data are needed. This includes the rather esoteric aerosol optical depth (AOD) and precipitable water vapor (PW) data required for most models.
3. Higher time resolution input data are required. AOD and PW data may sometimes be obtained for regions with similar climates but usually on a monthly or annual mean basis. Daily and hourly data will provide better, more natural time series data for more accurate system design and performance studies.

Regarding the models themselves, it is clear that little improvement has occurred since the mid-1990s. Most models discussed here are comparable in their performance. The need for more accurate models, or more accurate model uncertainties, requires the following:

1. Validation of all the broadband models described here (all were based on measured hourly average data) for various higher time resolution data, such as 10-min, 5-min, 1-min, and even 1-s "instantaneous" data.
2. Simple but accurate means of "extracting" high time resolution or small time step data from "summarized" data such as hourly average, daily average, or monthly average data (whether measured or modeled).
3. Development of a simpler, but more accurate clear sky model using less-esoteric (AOD, PW) input parameters that are easily (preferably universally) available.
4. Validation of the Perez anisotropic tilt conversion model over a wider range of climates and tilt configurations. The original model was only validated

for vertical tilts (90°) in the cardinal directions and 30° south-facing tilts. Refinements of the ε, Δ, and Fij parameters would improve the model.

5. Evaluation of the Maxwell DISC (direct insolation simulation code) GHI-to-DNI conversion model equations over a much wider range of latitudes, including a wider range of (low) zenith angles. The model is known to underpredict DNI at low zenith angles because of the limited data set (continental United States) used to develop the model equations.

6. Better, more accurate correlation, with smaller random error, relation between GHI and DHI irradiances. This supplements and complements an improved DISC model.

7. Better cloud cover modifiers, to somehow account for cloud spatial distributions with respect to the sun, either through (easily accessible, free) sky dome imagery, satellite imagery, or cloud forecast/hindcast data. This will result in better cloud transient modeling of solar radiation available to large-area (several-square-kilometer) systems.

As of 2012, many of these are current topics of research, especially for solar resource forecasting and future solar radiation database updates. Obtaining the financial, intellectual, and material (measurements) resources needed to meet these needs in difficult, or even less-stressful, economic times is a daunting task. But, given enough time, money, and computer power—at least that is the hope. I have the fervent belief, for the sake of future generations and the sustainability of the environment for them, that these goals, as well as the existing tools described in this book, can be powerful tools for success in fostering renewable energy systems of the future.

Appendix A: Bird Clear Sky Model in Excel

Excel® Column A holds user inputs for Bird clear sky model in rows 6 to 29. Note the cell references rely on the exact structure of rows and columns shown. User values are entered in shaded cells. Bold text are labels for values to be entered in cell below label.

TABLE A.1

Bird Clear Sky Input Parameters

Row Number	Column A, Label, Value, or Function
1	
2	
3	*N/A*
4	
5	
6	*USER*
7	*INPUTS*
8	Latitude
9	User Value (+N, –S)
10	Longitude
11	User Value (+E, –W)
12	Time Zone
13	User Value (–W, +E)
14	Station Pressure mB
15	User Value
16	Ozone atm-cm
17	User Value (0.0–0.35)
18	Water Vapor H$_2$O atm-cm
19	User Value (0.0–6.0)
20	Aerosol Optical Depth (AOD) @ 500 nm
21	User Value (0.0–0.4)
22	AOD @ 380 nm
23	User value (~1.5 × AOD @ 500 nm)
24	Taua (total AOD)
25	= 0.2758*$A$23+0.35*$A$21
26	Asymmetry Factor Ba
27	0.85
28	Ground Albedo (0.0 to 1.0)
29	0.2

TABLE A.2
Values and Functions to Be Entered in Row 3, from Columns B to V

Column	Row 1	Row 2	Rows 3–8762
B		DOY	1 to 365
C		HR	0 to 23
D	**Solar Position**	ETR	$= 1367*(1.00011+0.034221*COS(6.28318*(B3-1)/365)$ $+0.00128*SIN(6.28318*(B3-1)/365)+0.000719$ $*COS(2*(6.28318*(B3-1)/365))+0.000077$ $*SIN(2*(6.28318*(B3-1)/365)))$
E		Dangle	$= 6.283185*(B3-1)/365$
F		DEC	$= (0.006918-0.399912*COS(E3)+0.070257*SIN(E3)$ $-0.006758*COS(2*E3)+0.000907*SIN(2*E3)$ $-0.002697*COS(3*E3)+0.00148*SIN(3*E3))$ $*(180/3.14159)$
G		EQT	$= (0.000075+0.001868*COS(E3)-0.032077*SIN(E3)$ $-0.014615*COS(2*E3)-0.040849*SIN(2*E3))*(229.18)$
H		Hour Angle	$= 15*((C3-0.5)-\$A\$13-12)+\$A\$11+G3/4$
I		Zenith Angle	$= ACOS(COS(F3/(180/3.14159))*COS(\$A\$9/(180/3.14159))$ $*COS(H3/(180/3.14159))+SIN(F3/(180/3.14159))$ $*SIN(\$A\$9/(180/3.14159)))*(180/3.14159)$
J		Air Mass	$= IF(I3<89,1/(COS(I3/(180/3.14159))$ $+0.15/(93.885-I3)^1.25),0)$
K	Transmittances	T Rayleigh	$= IF(J3>0,EXP(-0.0903*(\$A\$15/1013)^0.84$ $*(1+\$A\$15/1013-(\$A\$15/1013)^1.01)),0)$
L		T Ozone	$= IF(J3>0,1-0.1611*(\$A\$17*J3)*(1+139.48*(\$A\$17*J3))^$ $-0.3034-0.002715*(\$A\$17*J3)/(1+0.044*(\$A\$17*J3)$ $+0.0003*(\$A\$17*J3)^2),0)$
M		T Gases	$= IF(J3>0,EXP(-0.0127*(J3*\$A\$15/1013)^0.26),0)$
N		T Water	$= IF(J3>0,1-2.4959*J3*\$A\$19/((1+79.034*\$A\$19*J3)$ $^0.6828+6.385*\$A\$19*J3),0)$
O		T Aerosol	$= IF(J3>0,EXP(-(\$A\$25^0.873)$ $*(1+\$A\$25-\$A\$25^0.7088)*J3^0.918),0)$
P		TAA	$= IF(J3>0,1-0.1*(1-J3+J3^1.06)*(1-O3),0)$
Q	Intermediate Results	Rs	$= IF(J3>0,0.0685+(1-\$A\$27)*(1-O3/P3),0)$
R		Id	$= IF(J3>0,0.9662*D3*O3*N3*M3*L3*K3,0)$
S		IdnH	$= IF(I3<90,R3*COS(I3/(180/3.14159)),0)$
T		Ias	$= IF(J3>0,D3*COS(I3/(180/3.14159))*0.79*L3*M3*N3*P3$ $*(0.5*(1-K3)+\$A\$27*(1-(O3/P3))/(1-J3+(J3)^1.02)),0)$
U		GH	$= IF(J3>0,(S3+T3)/(1-\$A\$29*Q3),0)$
V	Decimal	Time	$= B3+(C3-0.5)/24$
W	**MODEL RESULT** **(W/M²)**	Direct Beam	$= R3$
X		Direct Hz	$= S3$
Y		Global Hz	$= U3$
Z		Diffuse Hz	$= Y3-X3$

Note: Text in columns 2 and 3 are simply labels to be entered into designated cell row/column location. Column 3 contains functions with references to row 3 cell contents, to be copied to the rows following.

Appendix B: Excel Structure for DISC Model of Direct Normal Irradiance (DNI) from Global Horizontal Irradiance (GHI)

Tables B.1 and B.2 show the structure and functions for implementing the DISC (direct insolation simulation code) model in Excel®. Note the cell references rely on the exact structure of rows and columns shown. User values are entered in shaded cells.

TABLE B.1

User Location Input Parameters

Row	Column A Label/Value
1	
2	
3	
4	USER
5	INPUTS
6	**Latitude**
7	User Value (+N, –S)
8	**Longitude**
9	User Value (+E, –W)
10	**Time Zone**
11	User Value (–W, +E)
12	**Stn. Pressure**
13	User Value mB
14	**Pstn/P0**
15	= A13/1013.250

TABLE B.2
Functions to Be Entered in Row 2, from Columns G to W

Column	Label (Row 1)	Value or Function (Rows 2–8761)
B	Month	1 to 12
C	Day	1 to 28, 30, or 31
D	DOY	1 to 365
E	Hr	1 to 23
F	Obs. Hour	1 to 8760
G	ETR	= 1370*(1.00011+0.034221*COS(H2)+0.00128*SIN(H2)+0.000719 *COS(2*H2) +0.000077*(SIN(2*H2)))
H	Day Angle	= 6.283185*(D2–1)/365
I	DEC	= (0.006918–0.399912*COS(H2)+0.070257*SIN(H2)–0.006758 *COS(2*H2) +0.000907*SIN(2*H2)–0.002697 *COS(3*H2)+0.00148 *SIN(3*H2))*(180/3.14159)
J	EQT	= (0.000075+0.001868*COS(H2)–0.032077*SIN(H2)–0.014615 *COS(2*H2)–0.040849*SIN(2*H2))*(229.18)
K	Hour Angle	= 15*(E2–12.5+ J2/60 +(($A$9–$A$11*15)*4)/60)
L	Zenith Ang	= ACOS(COS(I2/(180/3.14159))*COS($A$7/(180/3.14159)) *COS(K2/(180/3.14159))+SIN(I2/(180/3.14159)) *SIN($A$7/(180/3.14159))) *(180/3.14159)
M	Air Mass	= IF(L2<89,$A$15*(1/(COS(L2/(180/3.14159)) +0.15/(93.885–L2)^1.253)),0)
N	Global Horizontal	USER VALUE
O	Diffuse (optional)	USER VALUE
P	Direct (Optional)	USER VALUE
Q	Kt	= IF(L2<80,N2/(COS(L2/(180/3.14159))*G2),0)
R	DISC Coeff. A	= IF(Q2>0,IF(Q2>0.6, –5.743+21.77*Q2–27.49*Q2^2+11.56 *Q2^3, IF(Q2<0.6,0.512–1.56*Q2+2.286*Q2^2–2.222*Q2^3)),0)
S	DISC Coeff. B	= IF(Q2>0,IF(Q2>0.6, 41.4–118.5*Q2+66.05*Q2^2+31.9 *Q2^3, IF(Q2<0.6,0.37+0.962*Q2,0)),0)
T	DISC Coeff. C	= IF(Q2>0,IF(Q2>0.6, –47.01+184.2*Q2–222*Q2^2+73.81 *Q2^3, IF(Q2<0.6,–0.28+0.932*Q2–2.048*Q2^2,0)),0)
U	DKn	= IF(Q2>0,(R2+S2*EXP(T2*M2))*$A$15,0)
V	Knc	= IF(Q2>0,(0.886–0.122*M2+0.0121*(M2)^2–0.000653 *(M2)^3+0.000014*(M2)^4),0)
W	Model DNI	= IF(Q2>0,G2*(V2–U2),0)

Note: Columns B to F may use either specific, noncontinuous values or be filled as serially complete to the extent needed. Column N is the required user value for global horizontal total hemispherical irradiance.

Appendix C: Tables for CIE D65 Reference Spectrum and Spectral Daylight Temperature Model

TABLE C.1

CIE D65 Reference Spectrum Relative Amplitude, S_0, S_1, S_2 Functions for Use in Equations 7.2–7.4

Wavelength (nm)	D65 Reference Spectrum S(λ) (Relative Units)	$S_0(\lambda)$	$S_1(\lambda)$	$S_2(\lambda)$	D50 Example Calculation (Relative Units)
300	0.03	0.04	0.02	0.00	0.02
305	1.66	3.02	2.26	1.00	1.04
310	3.29	6.00	4.50	2.00	2.05
315	11.77	17.80	13.45	3.00	4.92
320	20.24	29.60	22.40	4.00	7.78
325	28.64	42.45	32.20	6.25	11.27
330	37.05	55.30	42.00	8.50	14.76
335	38.50	56.30	41.30	8.15	16.36
340	40.00	57.30	40.60	7.80	17.96
345	42.43	59.50	41.10	7.25	19.44
350	44.91	61.80	41.60	6.70	21.02
355	45.78	61.65	39.80	6.00	22.49
360	46.64	61.50	38.00	5.30	23.95
365	49.36	65.15	40.20	5.70	25.46
370	52.09	68.80	42.40	6.10	26.97
375	51.03	66.10	40.45	4.55	25.74
380	49.98	63.40	38.50	3.00	24.50
385	52.31	64.60	36.75	2.10	27.19
390	54.65	65.80	35.00	1.20	29.88
395	68.70	80.30	39.20	0.05	39.60
400	82.75	94.80	43.40	−1.10	49.32
405	87.12	99.80	44.85	−0.50	53.03
410	91.49	104.80	46.30	−0.50	56.53
415	92.46	105.30	45.10	−0.60	58.24
420	93.43	105.90	43.90	−0.70	60.05

TABLE C.1 *(Continued)*
CIE D65 Reference Spectrum Relative Amplitude,
S_0, S_1, S_2 Functions for Use in Equations 7.2–7.4

Wavelength (nm)	D65 Reference Spectrum S(λ) (Relative Units)	$S_0(\lambda)$	$S_1(\lambda)$	$S_2(\lambda)$	D50 Example Calculation (Relative Units)
425	90.06	101.35	40.50	−0.95	58.94
430	86.68	96.80	37.10	−1.20	57.83
435	95.77	105.35	36.90	−1.95	66.32
440	104.86	113.90	36.70	−2.60	74.84
445	110.94	119.75	36.30	−2.75	81.05
450	117.01	125.60	35.90	−2.90	87.26
455	117.41	125.50	34.25	−2.85	88.89
460	117.81	125.50	32.60	−2.80	90.62
465	116.34	123.40	30.25	−2.70	91.00
470	114.86	121.30	27.90	−2.60	91.38
475	115.39	121.30	26.10	−2.60	93.25
480	115.92	121.30	24.30	−2.60	95.12
485	112.37	117.40	22.20	−2.20	93.54
490	108.81	113.50	20.10	−1.80	91.97
495	109.80	113.30	18.15	−1.66	93.85
500	109.35	113.10	16.20	−1.50	95.73
505	108.58	111.95	14.70	−1.40	96.17
510	107.80	110.80	13.20	−1.30	96.62
515	106.30	108.65	10.90	−1.25	96.88
520	104.79	106.50	8.60	−1.20	97.13
525	106.24	107.65	7.35	−1.10	99.62
530	107.69	108.80	6.10	−1.00	102.10
535	106.05	107.05	5.15	−0.75	101.43
540	104.41	105.30	4.20	−0.50	100.76
545	104.23	104.85	3.05	−0.30	101.57
550	104.05	104.40	1.90	−0.30	102.32
555	102.02	102.40	0.95	−0.15	101.36
560	100.00	100.00	0.00	0.00	100.00
565	98.17	98.00	−0.80	0.10	98.87
570	96.33	96.00	−1.60	0.20	97.73
575	96.06	95.55	−2.55	0.30	98.31
580	95.79	95.10	−3.50	0.50	98.92
585	92.24	92.10	−3.50	1.30	96.21
590	88.69	89.10	−3.50	2.10	93.50
595	89.35	89.80	−4.65	2.65	95.59
600	90.01	90.50	−5.80	3.20	97.68
605	89.80	90.40	−6.05	3.65	98.01
610	89.60	90.30	−7.20	4.10	99.27

TABLE C.1 *(Continued)*

CIE D65 Reference Spectrum Relative Amplitude,
S_0, S_1, S_2 Functions for Use in Equations 7.2–7.4

Wavelength (nm)	D65 Reference Spectrum $S(\lambda)$ (Relative Units)	$S_0(\lambda)$	$S_1(\lambda)$	$S_2(\lambda)$	D50 Example Calculation (Relative Units)
615	88.65	89.35	−7.90	4.40	99.15
620	87.70	88.40	−8.60	4.70	99.04
625	85.49	86.20	−9.05	4.90	97.38
630	83.29	84.00	−9.50	5.10	95.72
635	83.49	84.55	−10.20	5.90	97.28
640	83.70	85.10	−10.90	6.70	98.85
645	81.86	83.50	−10.80	7.00	97.26
650	80.03	81.90	−10.70	7.30	95.66
655	80.12	82.25	−11.35	8.00	96.94
660	80.21	82.60	−12.00	8.60	98.18
665	81.25	83.75	−13.00	9.20	100.59
670	82.28	84.90	−14.00	9.80	103.00
675	80.28	83.10	−13.80	10.00	101.06
680	78.28	81.30	−13.60	10.20	99.13
685	74.00	79.60	−12.80	9.25	96.25
690	69.72	71.90	−12.00	8.30	87.37
695	70.67	73.10	−12.65	8.95	89.49
700	71.61	74.30	−13.30	9.60	91.60
705	72.98	75.35	−13.10	9.08	92.25
710	74.35	76.40	−12.90	8.50	92.88
715	67.98	69.85	−11.75	7.75	84.87
720	61.60	63.30	−10.60	7.00	76.85
725	65.74	67.50	−11.10	7.30	81.68
730	69.89	71.70	−11.60	7.60	86.50
735	72.49	74.35	−11.90	7.80	89.54
740	75.09	77.00	−12.20	8.00	92.57
745	69.34	71.10	−11.20	7.35	85.40
750	63.59	65.20	−10.20	6.70	78.22
755	55.01	56.45	−9.00	5.95	67.96
760	46.42	47.70	−7.80	5.20	57.69
765	56.61	58.15	−9.50	6.30	70.30
770	66.81	68.60	−11.20	7.40	82.92
775	65.09	66.80	−10.80	7.10	80.59
780	63.38	65.00	−10.40	6.80	78.27
785	63.84	65.50	−10.50	6.90	78.91
790	64.30	66.00	−10.60	7.00	79.55
795	61.88	63.50	−10.15	6.70	76.47
800	59.45	61.00	−9.70	6.40	73.40

TABLE C.1 *(Continued)*
CIE D65 Reference Spectrum Relative Amplitude,
S_0, S_1, S_2 **Functions for Use in Equations 7.2–7.4**

Wavelength (nm)	D65 Reference Spectrum $S(\lambda)$ (Relative Units)	$S_0(\lambda)$	$S_1(\lambda)$	$S_2(\lambda)$	D50 Example Calculation (Relative Units)
805	55.71	57.15	−9.00	5.95	68.66
810	51.96	53.30	−8.30	5.50	63.92
815	54.70	53.30	−8.30	5.80	64.02
820	57.44	58.90	−9.30	6.10	70.77
825	58.88	60.40	−9.55	6.30	72.60
830	60.31	61.90	−9.80	6.50	74.44

Appendix D: SPCTRL2 FORTRAN Source Code

(+ is "continuation code" an for expression requiring multiple lines; "C" represents a comment line.)

```
PROGRAM SPCTRAL2 ( INPUT,OUTPUT,TAPE5=INPUT,TAPE6=OUTPUT,
    +TAPE1,TAPE2,TAPE3)
  C
  C...TAPE1 = INPUT PARAMETERS
  C...TAPE2 = EXTRATERRESTRIAL SPECTRAL IRRADIANCE, ABSORPTION COEFFICIENTS.
  C...TAPE3 = OUTPUT FILE
  C...TAPE5 = TERMINAL SCREEN
  C
  C...THIS CODE WAS WRITTEN BY RICHARD BIRD AND CALCULATES SPECTRAL
  C   IRRADIANCE OR PHOTON FLUX ON A TILTED OR HORIZONTAL SURFACE.
  C   MODIFICATIONS TO THE CODE TO CONVERT SPECTRAL IRRADIANCE TO PHOTON
  C   FLUX PER WAVELENGTH OR ELECTRON VOLT WERE MADE BY CAROL RIORDAN.
  C   COMMENTS AND PROGRAM INSTRUCTIONS WERE ADDED BY RIORDAN 8/84.
  C
  C   ******** INPUT VARIABLES ********
  C   AI    = ANGLE OF INCIDENCE OF DIRECT BEAM ON FLAT SURFACE (DEG)
  C   ALPHA = POWER ON ANGSTRON TURBIDITY EXPRESSION (1.14 FOR RURAL)
  C   NDAY  = JULIAN DAY
  C   NW    = NUMBER OF WAVELENGTHS FOR THIS RUN
  C   O3    = OZONE AMOUNT (ATM CM)
  C   RHO   = GROUND ALBEDO
  C   TILT  = TILT ANGLE OF SURFACE FROM THE HORIZONTAL (DEG)
  C   SPR   = SURFACE PRESSURE (MILLIBARS)
  C   TAU5  = THE AEROSOL OPTICAL DEPTH AT 0.5 MICRONS (BASE E)
  C   W     = PRECIPITABLE WATER VAPOR (CM)
  C   WV(I) & R(I) GROUND REFLECTIVITY AT SPECIFIC WAVELENGTHS
  C   Z     = SOLAR ZENITH ANGLE (DEG)
  C
  C   ***** TO RUN THIS PROGRAM *****
  C
  C...  IN THE FIRST LINE OF INPUT ON TAPE1, CHOOSE FROM THE
  C     FOLLOWING OPTIONS:
  C         COLLECTOR MODE
  C             (MODE) 1=GLOBAL NORMAL AND DIRECT NORMAL
  C                       (TRACKING FLAT PLATE/CONCENTRATOR)
  C                    2=GLOBAL TILT (FIXED FLAT PLATE)
  C                    3=GLOBAL HORIZONTAL
  C
  C         OUTPUT UNITS     (NUNITS) 1=IRRADIANCE PER WAVELENGTH
  C                                   2=PHOTON FLUX PER WAVELENGTH
  C                                   3=PHOTON FLUX PER ELECTRON VOLT
  C
  C         NUMBER OF SPECTRA (NUMSPT)
  C
  C   CHECK THE INPUT PARAMETERS ON THE SECOND LINE OF TAPE1, ESPECIALLY
  C   TAU5, W, TILT, SPR, NDAY.  Z AND AI ARE READ IN THE "DO 15 LOOP".
  C
  C   ******* THE PROGRAM *****
      DIMENSION WV(6),R(6)
  C
  C...READ INPUT
```

```
      READ(1,1,) MODE,NUNITS,NUMSPT
      READ(1,2) TAU5,ALPHA,O3,W,TILT,SPR,NDAY,NW
      READ(1,3) V(1),R(1),WV(2),R(2),WV(3),R(3),WV(4),R(4),WV(5),R(5),
     + WV(6),R(6)
C
C...THESE CONSTANTS ARE NEEDED TO CALCULATE PHOTON FLUX.
C
      XH=6.626176*(10.**(-34.))
      XC=2.99792458*(10.**8.)
      EVOLT=1.6021892*(10.**(-19.))
      CONST=(1./(XH*XC))*(10.**(-10.))
      XC=XC*(10.**6.)
C
C...RADIANS PER DEGREE
      RPD=.0174533
C
C...ERV IS THE EARTH RADIUS VECTOR-CORRECTION FOR SUN DISTANCE
      TT=6.283185*(NDAY-1)/365.
      TT2=2.*TT
      ERV=1.00011+.034221*COS(TT)+.00128*SIN(TT)
     +  + .000719*COS(TT2)+.000077*SIN(TT2)
C
C...WRITE OUT INITIAL CONDITIONS
      IF(MODE .EQ. 1) WRITE(5,20)
      IF(MODE .EQ. 2) WRITE(5,21)
      IF(MODE .EQ. 3) WRITE(5,22)
      WRITE(5,4) NUNITS,NUMSPT
      WRITE(5,5) NDAY,ERV
      WRITE(5,7) TAU5,ALPHA,W,O3
      WRITE(5,8) WV(1),R(1),WV(2),R(2),WV(3),R(3),WV(4),R(4),
     + WV(5),R(5),WV(6),R(6)
C
C...OMEG AND OMEGP ARE USED IN THE SINGLE SCATTERING ALBEDO
C    CALCULATIONS. JUSTUS HAS A FORM OF OMEGL THAT VARIES WITH
C    RELATIVE HUMIDITY AS WELL AS WAVELENGTH.
C
      OMEG=0.945
      OMEGP=0.095
C
C...FS AND FSP ARE THE FORWARD/TOTAL SCATTERING RATIOS AS
C    A FUNCTION OF ZENITH ANGLE.AIR MASS FOR FSP IS FIXED IN THIS CODE,
C    FS IS IN THE "DO 15 LOOP". GG IS THE AEROSOL ASSYMETRY
C    FACTOR (0.65 USED FOR RURAL).
C
      GG=0.65
      ALG=ALOG(1.-GG)
      AFS=ALG*(1.459+ALG*(0.1595+ALG*0.4129))
      BFS=ALG*(0.0783+ALG*(-0.3824-ALG*0.5874))
      FSP=1.-0.5*EXP((AFS+BFS/1.8)/1.8)
C
C...RR IS USED IN THE OZONE MASS EXPRESSION IN THE "DO 15 LOOP".
      RR=22./6370.
C
C...THESE EXPRESSIONS DEPEND ON ZENITH AND INCIDENCE ANGLE AND
C    SHOULD BE INSIDE THE LOOP THAT CHANGES Z AND AI.
C
C...FOR GLOBAL NORMAL SPECTRA (MODE=1), INCIDENT ANGLE(AI)=0 AND
C    TILT= Z. FOR GLOBAL HORIZONTAL SPECTRA
C    (MODE=3), INCIDENCE ANGLE (AI)=Z AND TILT=0.
C
C...DO LOOP FOR SEVERAL SPECTRA (NUMSPT)
      DO 15 KK=1, NUMSPT
           READ(1,10) Z, AI
           IF (MODE .EQ. 1) AI=0.
```

```
                IF (MODE .EQ. 1) TILT=Z
                IF (MODE .EQ. 3) TILT=0.
                IF (MODE .EQ. 3) AI=Z
                COSTLT=COS(TILT*RPD)
                CI=COS(AI*RPD)
                ZCOS=COS(Z*RPD)
                ZSIN=SIN(Z*RPD)
                FS=1.-0.5*EXP((AFS+BFS*ZCOS)*ZCOS)
C
C...RELATIVE OPTICAL AIR MASS.
                AM=1./(ZCOS+.15*(93.885-Z)**(-1.253))
C...PRESSURE CORRECTED AIR MASS.
                AMP=AM*SPR/1013.
C
C...OZONE MASS.
                AMO=(1.+RR)/(ZCOS**2.+2.*RR)**.5
C
C...WRITE OUT INITAL CONDITIONS TO TTERMINAL SCREEN
C
        WRITE(5,17) Z,TILT,AI
        WRITE(5,6) AM,SPR,AMP,AMO
C
C...INITIALIZE THE INTERPOLATION COUNTER FOR RHO.
        NR=2
C...REWIND THE TAPE WITH ET SPECTRUM AND ABSORPTION COEFFICIENTS FOR
C   EACH SPECTRUM.
        REWIND 2
C
C...DO LOOP FOR NUMBER OF WAVELENGTHS (NW)
        DO 14 I=1,NW
                READ(2,9) WVL,H0,AW,AO,AU
C
C...CORRECT EXTRATERRESTRIAL IRRADIANCE FOR EARTH RADIUS VECTOR.
                H0=H0*ERV
C
C
C...OMEGL IS THE WAVELENGTH-DEPENDENT SINGLE SCATTERING ALBEDO
C
                OMEGL=OMEG*EXP(-OMEGP*(ALOG(WVL/0.4))**2.)
C
C...THE FOLLOWING STATEMENTS PRODUCE THE WAVELENGTH-DEPENDENT ALBEDO
C
                IF(WVL .GT. WV(NR)) NR=NR+1
                SLP=(R(NR)-R(NR-1))/(WV(NR)-WV(NR-1))
                RHO=SLP*(WVL-WV(NR-1))+R(NR-1)
C
C...CALCULATE TRANSMITTANCE
                TR=EXP(-AMP/(WVL**4.*(115.6406-1.335/WVL**2.)))
                TO=EXP(-AO*O3*AMO)
                TW=EXP(-.2385*AW*W*AM/(1.+20.07*AW*W*AM)**.45)
                TU=EXP(-1.41*AU*AMP/((1.+118.93*AU*AMP)**.45))
                DELA=TAU5*(WVL/.5)**(-ALPHA)
                TAS=EXP(-OMEGL*DELA*AM)
                TAA=EXP(-(1.-OMEGL)*DELA*AM)
                TA=EXP(-DELA*AM)
C
                DIR=H0*TR*TO*TW*TU*TA
C...CALCULATE DIRECT COMPONENT OF IRRADIANCE ONTO SURFACE
                DIRSUR=DIR*CI
C
C...DRAY AND DAER HAVE BEEN MODIFIED BY BIRD. NOTE POWER ON TR TERM.
C
                DRAY=H0*ZCOS*TO*TW*TU*TAA*(1.-TR**.95)*.5
                DAER=H0*ZCOS*TO*TW*TU*TAA*TR**1.5*(1.-TAS)*FS
```

```
          TRP=EXP(-1.8/(WVL**4.*(115.6406-1.335/WVL**2)))
          TWP=EXP(-.2385*AW*W*1.8/((1.+20.07*AW*W*1.8)**.45))
          TUP=EXP(-1.41*AU*1.8/((1.+118.93*AU*1.8)**.45))
          TASP=EXP(-OMEGL*DELA*1.8)
          TAAP=EXP(-(1.-OMEGL)*DELA*1.8)
          RHOA=TUP*TWP*TAAP*(.5*(1.-TRP)+(1.-FSP)*TRP*(1.-TASP))
          DRGD=(DIR*ZCOS+(DRAY+DAER))*RHO*RHOA/(1.-RHO*RHOA)
          DIF=DRAY+DAER+DRGD
C
C...CRC IS A UV CORRECTION FACTOR
C
          CRC=1.0
          IF (WVL .LE. .45) CRC=(WVL+.55)**1.8
C
C...DIFFUSE ON A HORIZONTAL SURFACE
          DIF=DIF*CRC
C
C...TOTAL ON A HORIZONTAL SURFACE
          DTOT=DIR*ZCOS+DIF
C...MAKE DIFS=DIF TO USE FORMAT 12 FOR GLOBAL HORIZONTAL CASE.
          DIFS=DIF
          IF (MODE .EQ. 3) GOTO 79
C
C...GROUND REFLECTED COMPONENT
          REFS=DTOT*RHO*(1.0-COSTLT)/2.0
C
C...THE THREE FOLLOWING STATEMENTS ARE THE HAY TILT ALGORITHM
C
C...ANISOTROPY INDEX
          AII=DIR/H0
C
C...CIRCUMSOLAR AND ISOTROPIC COMPONENT (WEIGHTED BY DIF*AII AND
C   DIF*(1-AII).
          DIFSC=DIF*AII*CI/ZCOS
          DIFSI=DIF*(1.0-AII)*(1.0+COSTLT)/2.0
C
C...DIFFUSE ON TILTED SURFACE.
          DIFS=DIFSC+DIFSI+REFS
C
C...TOTAL ON TILTED SURFACE.
          DTOT=DIR*CI+DIFS


C
C...PHOTON FLUX CALCULATIONS.
C
C...WRITE SPECTRAL IRRADIANCE OR PHOTON FLUX OUTPUT TO TAPE3.
   79     IF (NUNITS .EQ.1) GOTO 12
          PFWVGL=DTOT*WVL*CONST
          ENERGY=(XH*XC)/WVL
          E=ENERGY/EVOLT
          PFEVGL=(PFWVGL*WVL)/E
          PFWVDN=DIR*WVL*CONST
          PFEVDN=(PFWVDN*WVL)/E
C         PFWVDS=DIRSUR*WV*CONST
C         PFEVDS=(PFWVDS*WVL)/E
C.... DECIDE IF USING DIF OR DIFS
C         PFWVDF=DIF*WVL*CONST
C         PFEVDF=(PFWVDF*WVL)/E
          IF (NUNITS .EQ. 2) GOTO 11
          WRITE (3,18) E,PFEVGL,PFEVDN
          GOTO 13
   11     WRITE (3,18) WVL,PFWVGL,PFWVDN
          GOTO 13
   12 WRITE (3,16) WVL,DTOT,DIR,DIFS
   13 CONTINUE
```

```
C
 14   CONTINUE
 15   CONTINUE
C
C****** FORMAT STEMENTS *****
C
  1 FORMAT(3I5)
  2 FORMAT(6F6.3,2I5)
  3 FORMAT(12F6.2)
  4 FORMAT(2X,"NUNITS=",I6,3X,"NUMSPT=",I6/)
  5 FORMAT(2X,"NDAY =",I7,3X,ERV   =",F7.4)
  6 FORMAT(2X,"AM     =",F7.2,3X,"SPR  =",F7.2,3X,"AMP  =",F7.2,3X,
    +"AMO  =",F7.2)
  7 FORMAT(2X,"TAU5 =",F7.2,3X,"ALPHA=",F7.2,3X,"W    =",F7.2,
    +3X,"O3   =",F7.3)
  8 FORMAT(2X,"W1=",F6.2,2X,"R1=",F6.2/2CX,"W2=",F6.2,2X,
    +"R2=",F6.2/2X,"W3=",F6.2,2X,"R3=",F6.2/2X,"W4=",F6.2,2X,
    +"R4=",F6.2/2X,"W5=",F6.2,2X,"R5=",F6.2/2X,"W6=",F6.2,2X,
    +"R6=",F6.2/)
  9 FORMAT(5F10.4)
 10 FORMAT(2F10.4)
 16 FORMAT(F7.4,3F10.4)
 17 FORMAT(2X,"Z    =",F7.2,3X,"TILT =",F7.2,3X,"AI   =",F7.2)
 18 FORMAT(F10.5,2E10.4)
 20 FORMAT(1X,"GLOBAL NORMAL AND DIRECT NORMAL")
 21 FORMAT(1X,"GLOBAL TILT")
 22 FORMAT(1X,"GLOBAL HORIZONTAL")
    STOP
    END
```

TABLE D.1
Fixed Spectral Data for SPCTRL2 Clear Sky Model

Wavelength Microns (υm)	Wehrli ETR W/cm²/υm	Water Absorption	O3 Absorption	Mixed Gases
0.300	535.9	0	10	0
0.305	558.3	0	4.8	0
0.310	622	0	2.7	0
0.315	692.7	0	1.35	0
0.320	715.1	0	0.8	0
0.325	832.9	0	0.38	0
0.330	961.9	0	0.16	0
0.335	931.9	0	0.075	0
0.340	900.6	0	0.04	0
0.345	911.3	0	0.019	0
0.350	975.5	0	0.007	0
0.360	975.9	0	0	0
0.370	1119.9	0	0	0
0.380	1103.8	0	0	0
0.390	1033.8	0	0	0
0.400	1479.1	0	0	0
0.410	1701.3	0	0	0
0.420	1740.4	0	0	0
0.430	1587.2	0	0	0

TABLE D.1 *(Continued)*
Fixed Spectral Data for SPCTRL2 Clear Sky Model

Wavelength Microns (ʋm)	Wehrli ETR W/cm²/ʋm	Water Absorption	O3 Absorption	Mixed Gases
0.440	1837	0	0	0
0.450	2005	0	0.003	0
0.460	2043	0	0.006	0
0.470	1987	0	0.009	0
0.480	2027	0	0.014	0
0.490	1896	0	0.021	0
0.500	1909	0	0.03	0
0.510	1927	0	0.04	0
0.520	1831	0	0.048	0
0.530	1891	0	0.063	0
0.540	1898	0	0.075	0
0.550	1892	0	0.085	0
0.570	1840	0	0.12	0
0.593	1768	0.075	0.119	0
0.610	1728	0	0.12	0
0.630	1658	0	0.09	0
0.656	1524	0	0.065	0
0.668	1531	0	0.051	0
0.690	1420	0.016	0.028	0.15
0.710	1399	0.0125	0.018	0
0.718	1374	1.8	0.015	0
0.724	1373	2.5	0.012	0
0.740	1298	0.061	0.01	0
0.753	1269	0.0008	0.008	0
0.758	1245	0.0001	0.007	0
0.763	1223	0.00001	0.006	4
0.768	1205	0.00001	0.005	0.35
0.780	1183	0.0006	0	0
0.800	1148	0.036	0	0
0.816	1091	1.6	0	0
0.824	1062	2.5	0	0
0.832	1038	0.5	0	0
0.840	1022	0.155	0	0
0.860	998.7	0.00001	0	0
0.880	947.2	0.0026	0	0
0.905	893.2	7	0	0
0.915	868.2	5	0	0
0.925	829.7	5	0	0
0.930	830.3	27	0	0
0.937	814	55	0	0

TABLE D.1 *(Continued)*
Fixed Spectral Data for SPCTRL2 Clear Sky Model

Wavelength Microns (υm)	Wehrli ETR W/cm²/υm	Water Absorption	O3 Absorption	Mixed Gases
0.948	786.9	45	0	0
0.965	768.3	4	0	0
0.980	767	1.48	0	0
0.994	757.6	0.1	0	0
1.040	688.1	0.00001	0	0
1.070	640.7	0.001	0	0
1.100	606.2	3.2	0	0
1.120	585.9	115	0	0
1.130	570.2	70	0	0
1.145	564.1	75	0	0
1.161	544.2	10	0	0
1.170	533.4	5	0	0
1.200	501.6	2	0	0
1.240	477.5	0.002	0	0.05
1.270	442.7	0.002	0	0.3
1.290	440	0.1	0	0.02
1.320	416.8	4	0	0.0002
1.350	391.4	200	0	0.00011
1.395	358.9	1000	0	0.00001
1.443	327.5	185	0	0.05
1.463	317.5	80	0	0.011
1.477	307.3	80	0	0.005
1.497	300.4	12	0	0.0006
1.520	292.8	0.16	0	0
1.539	275.5	0.002	0	0.005
1.558	272.1	0.0005	0	0.13
1.578	259.3	0.0001	0	0.04
1.592	246.9	0.00001	0	0.06
1.610	244	0.0001	0	0.13
1.630	243.5	0.001	0	0.001
1.646	234.8	0.01	0	0.0014
1.678	220.5	0.036	0	0.0001
1.740	190.8	1.1	0	0.00001
1.800	171.1	130	0	0.00001
1.860	144.5	1000	0	0.0001
1.920	135.7	500	0	0.001
1.960	123	100	0	4.3
1.985	123.8	4	0	0.2
2.005	113	2.9	0	21
2.035	108.5	1	0	0.13

TABLE D.1 *(Continued)*
Fixed Spectral Data for SPCTRL2 Clear Sky Model

Wavelength Microns (ʋm)	Wehrli ETR W/cm²/ʋm	Water Absorption	O3 Absorption	Mixed Gases
2.065	97.5	0.4	0	1
2.100	92.4	0.22	0	0.08
2.148	82.4	0.25	0	0.001
2.198	74.6	0.33	0	0.00038
2.270	68.3	0.5	0	0.001
2.360	63.8	4	0	0.0005
2.450	49.5	80	0	0.00015
2.500	48.5	310	0	0.00014
2.600	38.6	15000	0	0.00066
2.700	36.6	22000	0	100
2.800	32	8000	0	150
2.900	28.1	650	0	0.13
3.000	24.8	240	0	0.0095
3.100	22.1	230	0	0.001
3.200	19.6	100	0	0.8
3.300	17.5	120	0	1.9
3.400	15.7	19.5	0	1.3
3.500	14.1	3.6	0	0.075
3.600	12.7	3.1	0	0.01
3.700	11.5	2.5	0	0.00195
3.800	10.4	1.4	0	0.004
3.900	9.5	0.17	0	0.29
4.000	8.6	0.0045	0	0.025

Appendix E: Photopic Response Function V(λ) Curve

Lambda (nm)	Vlambda	Lambda (nm)	Vlambda	Lambda (nm)	Vlambda	Lambda (nm)	Vlambda
375.0	0.0000	394.5	0.0015	428.0	0.0255	467.0	0.0803
375.5	0.0000	395.0	0.0015	429.0	0.0264	468.0	0.0837
376.0	0.0000	395.5	0.0017	430.0	0.0273	469.0	0.0872
376.5	0.0000	396.0	0.0018	431.0	0.0283	470.0	0.0910
377.0	0.0000	396.5	0.0018	432.0	0.0294	471.0	0.0949
377.5	0.0000	397.0	0.0019	433.0	0.0304	472.0	0.0990
378.0	0.0000	397.5	0.0020	434.0	0.0315	473.0	0.1034
378.5	0.0000	398.0	0.0022	435.0	0.0326	474.0	0.1079
379.0	0.0000	398.5	0.0024	436.0	0.0337	475.0	0.1126
379.5	0.0001	399.0	0.0025	437.0	0.0347	476.0	0.1175
380.0	0.0002	399.5	0.0027	438.0	0.0358	477.0	0.1227
80.5	0.0002	400.0	0.0028	439.0	0.0369	478.0	0.1280
381.0	0.0002	401.0	0.0031	440.0	0.0379	479.0	0.1335
381.5	0.0002	402.0	0.0035	441.0	0.0388	480.0	0.1390
382.0	0.0003	403.0	0.0038	442.0	0.0398	481.0	0.1447
382.5	0.0003	404.0	0.0042	443.0	0.0406	482.0	0.1505
383.0	0.0003	405.0	0.0047	444.0	0.0415	483.0	0.1565
383.5	0.0003	406.0	0.0051	445.0	0.0424	484.0	0.1627
384.0	0.0003	407.0	0.0056	446.0	0.0433	485.0	0.1693
384.5	0.0004	408.0	0.0062	447.0	0.0441	486.0	0.1762
385.0	0.0004	409.0	0.0068	448.0	0.0450	487.0	0.1836
385.5	0.0004	410.0	0.0074	449.0	0.0459	488.0	0.1913
386.0	0.0005	411.0	0.0081	450.0	0.0468	489.0	0.1994
386.5	0.0005	412.0	0.0090	451.0	0.0477	490.0	0.2080
387.0	0.0005	413.0	0.0098	452.0	0.0487	491.0	0.2171
387.5	0.0006	414.0	0.0108	453.0	0.0498	492.0	0.2267
388.0	0.0006	415.0	0.0118	454.0	0.0509	493.0	0.2369
388.5	0.0006	416.0	0.0128	455.0	0.0521	494.0	0.2475
389.0	0.0007	417.0	0.0140	456.0	0.0534	495.0	0.2586
389.5	0.0007	418.0	0.0151	457.0	0.0549	496.0	0.2702
390.0	0.0008	419.0	0.0163	458.0	0.0564	497.0	0.2823
390.5	0.0009	420.0	0.0175	459.0	0.0581	498.0	0.2951
391.0	0.0009	421.0	0.0186	460.0	0.0600	499.0	0.3086
391.5	0.0010	422.0	0.0196	461.0	0.0626	500.0	0.3230
392.0	0.0010	423.0	0.0207	462.0	0.0653	501.0	0.3384
392.5	0.0011	424.0	0.0217	463.0	0.0680	502.0	0.3547

Lambda (nm)	Vlambda	Lambda (nm)	Vlambda	Lambda (nm)	Vlambda	Lambda (nm)	Vlambda
393.0	0.0012	425.0	0.0227	464.0	0.0709	503.0	0.3717
393.5	0.0013	426.0	0.0236	465.0	0.0739	504.0	0.3893
394.0	0.0014	427.0	0.0246	466.0	0.0770	505.0	0.4073
506.0	0.4256	546.0	0.9841	586.0	0.8048	626.0	0.3093
507.0	0.4443	547.0	0.9874	587.0	0.7931	627.0	0.2979
508.0	0.4634	548.0	0.9903	588.0	0.7812	628.0	0.2866
509.0	0.4829	549.0	0.9928	589.0	0.7692	629.0	0.2756
510.0	0.5030	550.0	0.9950	590.0	0.7570	630.0	0.2650
511.0	0.5236	551.0	0.9967	591.0	0.7448	631.0	0.2548
512.0	0.5445	552.0	0.9981	592.0	0.7324	632.0	0.2449
513.0	0.5657	553.0	0.9991	593.0	0.7200	633.0	0.2353
514.0	0.5870	554.0	0.9997	594.0	0.7075	634.0	0.2261
515.0	0.6082	555.0	1.0000	595.0	0.6949	635.0	0.2170
516.0	0.6293	556.0	0.9999	596.0	0.6822	636.0	0.2082
517.0	0.6503	557.0	0.9993	597.0	0.6695	637.0	0.1995
518.0	0.6709	558.0	0.9983	598.0	0.6567	638.0	0.1912
519.0	0.6908	559.0	0.9969	599.0	0.6438	639.0	0.1830
520.0	0.7100	560.0	0.9950	600.0	0.6310	640.0	0.1750
521.0	0.7282	561.0	0.9926	601.0	0.6182	641.0	0.1672
522.0	0.7455	562.0	0.9897	602.0	0.6053	642.0	0.1596
523.0	0.7620	563.0	0.9864	603.0	0.5925	643.0	0.1523
524.0	0.7778	564.0	0.9827	604.0	0.5796	644.0	0.1451
525.0	0.7932	565.0	0.9786	605.0	0.5668	645.0	0.1382
526.0	0.8081	566.0	0.9741	606.0	0.5540	646.0	0.1315
527.0	0.8225	567.0	0.9692	607.0	0.5411	647.0	0.1250
528.0	0.8363	568.0	0.9639	608.0	0.5284	648.0	0.1188
529.0	0.8495	569.0	0.9581	609.0	0.5156	649.0	0.1128
530.0	0.8620	570.0	0.9520	610.0	0.5030	650.0	0.1070
531.0	0.8738	571.0	0.9455	611.0	0.4905	651.0	0.1015
532.0	0.8850	572.0	0.9385	612.0	0.4780	652.0	0.0962
533.0	0.8955	573.0	0.9312	613.0	0.4657	653.0	0.0911
534.0	0.9054	574.0	0.9235	614.0	0.4534	654.0	0.0863
535.0	0.9149	575.0	0.9154	615.0	0.4412	655.0	0.0816
536.0	0.9237	576.0	0.9070	616.0	0.4291	656.0	0.0771
537.0	0.9321	577.0	0.8983	617.0	0.4170	657.0	0.0728
538.0	0.9399	578.0	0.8892	618.0	0.4050	658.0	0.0687
539.0	0.9472	579.0	0.8798	619.0	0.3930	659.0	0.0648
540.0	0.9540	580.0	0.8700	620.0	0.3810	660.0	0.0610
541.0	0.9603	581.0	0.8599	621.0	0.3689	661.0	0.0574
542.0	0.9660	582.0	0.8494	622.0	0.3568	662.0	0.0540
543.0	0.9713	583.0	0.8386	623.0	0.3448	663.0	0.0507
544.0	0.9760	584.0	0.8276	624.0	0.3328	664.0	0.0475
545.0	0.9803	585.0	0.8163	625.0	0.3210	665.0	0.0446

Lambda (nm)	Vlambda	Lambda (nm)	Vlambda	Lambda (nm)	Vlambda	Lambda (nm)	Vlambda
666.0	0.0418	706.0	0.0027	746.0	0.0002		
667.0	0.0391	707.0	0.0026	747.0	0.0001		
668.0	0.0366	708.0	0.0024	748.0	0.0001		
669.0	0.0342	709.0	0.0022	749.0	0.0001		
670.0	0.0320	710.0	0.0021	750.0	0.0001		
671.0	0.0300	711.0	0.0020	751.0	0.0001		
672.0	0.0281	712.0	0.0018	752.0	0.0001		
673.0	0.0263	713.0	0.0017	753.0	0.0001		
674.0	0.0247	714.0	0.0016	754.0	0.0001		
675.0	0.0232	715.0	0.0015	755.0	0.0001		
676.0	0.0218	716.0	0.0014	756.0	0.0001		
677.0	0.0205	717.0	0.0013	757.0	0.0001		
678.0	0.0193	718.0	0.0012	758.0	0.0001		
679.0	0.0181	719.0	0.0011	759.0	0.0001		
680.0	0.0170	720.0	0.0010	760.0	0.0001		
681.0	0.0159	721.0	0.0010	761.0	0.0001		
682.0	0.0148	722.0	0.0009	762.0	0.0001		
683.0	0.0138	723.0	0.0009	763.0	0.0000		
684.0	0.0128	724.0	0.0008	764.0	0.0000		
685.0	0.0119	725.0	0.0007	765.0	0.0000		
686.0	0.0111	726.0	0.0007	766.0	0.0000		
687.0	0.0103	727.0	0.0006	767.0	0.0000		
688.0	0.0095	728.0	0.0006	768.0	0.0000		
689.0	0.0088	729.0	0.0006	769.0	0.0000		
690.0	0.0082	730.0	0.0005	770.0	0.0000		
691.0	0.0076	731.0	0.0005	771.0	0.0000		
692.0	0.0071	732.0	0.0005	772.0	0.0000		
693.0	0.0066	733.0	0.0004	773.0	0.0000		
694.0	0.0061	734.0	0.0004	774.0	0.0000		
695.0	0.0057	735.0	0.0004	775.0	0.0000		
696.0	0.0053	736.0	0.0003	776.0	0.0000		
697.0	0.0050	737.0	0.0003	777.0	0.0000		
698.0	0.0047	738.0	0.0003	778.0	0.0000		
699.0	0.0044	739.0	0.0003	779.0	0.0000		
700.0	0.0041	740.0	0.0002	780.0	0.0000		
701.0	0.0038	741.0	0.0002	781.0	0.0000		
702.0	0.0036	742.0	0.0002	782.0	0.0000		
703.0	0.0034	743.0	0.0002	783.0	0.0000		
704.0	0.0031	744.0	0.0002	784.0	0.0000		
705.0	0.0029	745.0	0.0002	785.0	0.0000		

Appendix F: Perez Anisotropic Model Coefficients for Luminous Efficacy and Zenith Luminance Model

TABLE F.1

Efficacy and Zenith Luminance Coefficients for Each Sky Condition Bin

E bin	Global Luminous Efficacy				Direct Luminous Efficacy			
	a1	b1	c1	d1	a1	b1	c1	d1
1	96.63	−0.47	11.50	−9.16	37.20	−4.55	−2.98	117.12
2	107.45	0.79	1.79	−1.19	98.99	−3.46	−1.21	12.38
3	98.73	0.70	4.40	−6.95	109.83	−4.90	−1.71	−8.81
4	92.72	0.56	8.36	−8.31	110.34	−5.84	−1.99	−4.56
5	86.73	0.98	7.10	−10.94	106.36	−3.97	−1.75	−6.16
6	88.34	1.39	6.06	−7.60	107.19	−1.25	−1.51	−26.73
7	78.63	1.47	4.93	−11.37	105.75	0.77	−1.26	−34.44
8	99.65	1.86	−4.46	−3.15	101.18	1.58	−1.10	−8.29

E bin	Diffuse Luminous Efficacy				Zenith Luminance Value			
	a1	b1	c1	d1	a1	b1	c1	d1
1	97.24	−0.46	12.00	−8.91	40.86	26.77	−29.59	−45.75
2	107.22	1.15	0.59	−3.95	26.58	17.73	58.46	−21.25
3	104.97	2.96	−5.53	−8.77	19.34	2.28	100.00	0.25
4	102.39	5.59	−13.95	−13.90	13.25	−1.39	124.79	15.66
5	100.71	5.94	−22.75	−23.74	14.47	−5.09	160.09	9.13
6	106.42	3.83	−36.15	−28.83	19.76	−3.88	154.61	−19.21
7	141.88	1.90	−53.24	−14.03	28.39	−9.67	151.58	−69.39
8	152.23	0.35	−45.27	−7.98	42.91	−10.62	130.80	−164.08

Source: Perez, R., P. Ineichen, R. Seals, J. Michalsky, and R. Stewart. (1990). Modeling daylight availability and irradiance components from direct and global irradiance. *Solar Energy*, Vol. 44, No, 5, pp. 271–289. With permission from Elsevier.

TABLE F.2
Sky Condition Bin and High-Low
Boundaries for Bin

ε Bin	ε Low	ε High
1	1.000	1.065
2	1.065	1.230
3	1.230	1.500
4	1.500	1.950
5	1.950	2.800
6	2.800	4.500
7	4.500	6.200
8	6.200	…

Source: Perez et al. *The Development and Verification of the Perez Radiation Model.* Contractor Report Sand88-7030 National Laboratories. http://prod.sandia.gov/techlib/access-control.cgi/1988/887030.pdf.
ε is the ratio of diffuse/global irradiance.

TABLE F.3
Regression Equations for ai … di Perez Luminance Model Coefficients as Function of Sky Condition BIN Number (= x = 1 to 8)

	ai	bi	ci	di
Global	$0.1338x^5 - 3.0027x^4$ $+ 25.477x^3 - 100.56x^2$ $+ 175.78x - 1.34$	$0.006x^5 - 0.1463x^4$ $+ 1.3416x^3 - 5.6664x^2$ $+ 10.91x - 6.92$	$-0.056x^5 + 1.3253x^4$ $- 11.977x^3 + 50.509x^2$ $- 95.86x + 67.608$	$0.0723x^5 - 1.6774x^4$ $+ 14.665x^3 - 59.211x^2$ $+ 105.86x - 68.888$
Diffuse	$-0.1323x^5 + 2.6945x^4$ $- 19.092x^3 + 56.594x^2$ $- 64.726x + 122.29$	$0.0027x^5 - 0.0203x^4$ $- 0.2396x^3 + 2.2606x^2$ $- 3.5478x + 1.13$	$0.0592x^5 - 1.0984x^4 +$ $7.125x^3 - 19.505x^2$ $+ 12.779x + 12.49$	$-0.0802x^5 + 1.7002x^4$ $- 12.518x^3 + 37.826x^2$ $- 45.653x + 10.105$
Direct	$-0.3728x^4 + 7.8336x^3$ $- 59.026x^2 + 186.46x$ $- 97.001$	$-0.0527x^4 + 0.925x^3$ $- 5.1477x^2 + 10.482x$ $- 10.661$	$0.0068x^5 - 0.173x^4$ $+ 1.652x^3 - 7.2701x^2$ $+ 14.359x - 11.548$	$1.0026x^4 - 19.343x^3$ $+ 131.86x^2 - 378.97x$ $+ 382.3$
Zenith	$2.2944x^2 - 20.342x$ $+ 58.725$	$-0.1781x^3 + 3.3461x^2$ $- 22.662x + 47.514$	$-7.5579x^2 + 89.3x$ $- 102.78$	$-1.0904x^3 + 5.0388x^2$ $+ 17.48x - 67.334$

Source: Data from http://prod.sandia.gov/techlib/access-control.cgi/1988/887030.pdf accessed 27 Oct 2012.

Index

Printed and bound by CPI Group (UK) Ltd, Croydon, CR0 4YY

18/10/2024

01776273-0001